Astronomical observations

ASTRONOMICAL OBSERVATIONS

an optical perspective

GORDON WALKER

Department of Geophysics and Astronomy
University of British Columbia

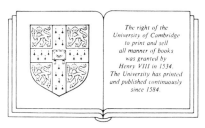

CAMBRIDGE UNIVERSITY PRESS

Cambridge

New York   New Rochelle

Melbourne   Sydney

Published by the Press Syndicate of the University of Cambridge
The Pitt Building, Trumpington Street, Cambridge CB2 1RP
32 East 57th Street, New York, NY 10022, USA
10 Stamford Road, Oakleigh, Melbourne 3166, Australia

First published 1987
Reprinted with corrections 1987

Printed and bound in Great Britain by
Redwood Burn Limited, Trowbridge, Wiltshire

British Library cataloguing in publication data

Walker, Gordon
Astronomical observations: an optical
perspective
1. Astronomical instruments
I. Title
522′.2   QB86

Library of Congress cataloging-in-publication data

Walker, Gordon Arthur Hunter, 1936–
Astronomical observations.

Bibliography
Includes index.
1. Astronomical instruments. 2. Astronomy – Technique.
3. Optical instruments. 4. Astrophysics. I. Title.
QB86.W35   1987   522   86-6833

ISBN 0 521 32587 0 hard covers
ISBN 0 521 33907 3 paperback

CONTENTS

For my parents

PREFACE

For a number of years I taught an undergraduate course on astronomical instrumentation and techniques. As there was never a very suitable textbook available I decided to write one myself. In it I have tried to demonstrate something of the unity and fundamental limitations of observational techniques across the electromagnetic spectrum while concentrating on the optical region which is the one most familiar to me. I assume that the reader has a sound background in physics, particularly optics, and is aware of recent discoveries and theories in astrophysics.

There is no attempt to give an exhaustive survey of existing instruments. Those I selected are illustrative of the techniques while others are covered in the references. Much of my material has been drawn from papers and reviews and, as far as possible, I have tried to acknowledge this by reference rather than by name in the text.

Most of the book was written in 1981–82 with support from a UBC Killam Senior Fellowship. Encouraged by colleagues, I completed the manuscript in the summer of 1985. This deadline led to some important omissions such as a critical discussion of calibration and astrometric techniques. Without the enormous help and enthusiasm of Steven Laker who, among other things, transformed all of the equations to SI units, I would never have finished at all. John Nicol, Stephenson Yang and Gerry Grieve gave me invaluable technical assistance in preparing the text and diagrams and computing funds were provided through an operating grant from the Canadian Natural Sciences and Engineering Research Council.

Many people and institutions supplied me with information and material for which I am very grateful. Among those not acknowledged in the text are K. Bennett of ESA and Luis Kaluzienski of NASA who provided the material for Tables 3.2 and 3.3, respectively. T. Landecker of DRAO and W. McCutcheon of UBC provided me with valuable references on radio astronomy. J. Krelowski made helpful comments on the manuscript.

Figures 1.2(c), 3.5, 4.7, 6.1, 6.23, 6.24, 6.25 are reprinted from the *Astrophysical Journal*, copyright © University of Chicago Press.

Gordon Walker
September 1986

1

Astronomical sources

The electromagnetic spectrum

All of our information about the universe beyond the solar system has, so far, come from electromagnetic radiation and cosmic rays. Early measurements were of the apparent position and brightness of sources and the angular motions of the nearer stars. With a better understanding of the nature of electromagnetic radiation, it became possible to estimate distances to objects which were too distant for a direct measurement of parallax and to describe the physical conditions in individual sources. In the past twenty-five years the quality of observations has improved dramatically as a result of instrumental developments, with astronomers having access to virtually the whole of the electromagnetic spectrum from telescopes on the ground or in space [135].

In 1690 Christian Huygens published the *Traité de la lumière* in which he proposed that any point source of light propagates a spherical wave into the surrounding medium with the velocity of light. He assumed that, at any instant, each point on the surface of the spherical wavefront would act as an independent source (with an appropriate obliquity factor) and he was able to establish a consistent interpretation for both geometrical optics and the diffraction of light.

In 1864 Clerk Maxwell laid the foundation of modern electromagnetic theory in a paper entitled 'A dynamical theory of the electromagnetic field'. He related the time and spatial derivatives of the electric and magnetic vectors in four equations. He demonstrated that electromagnetic waves could propagate in a vacuum and that they would involve only transverse electric and magnetic vibrations. In Figure 1.1 the relative phases and orientation of the $\mathbf{E}$ and $\mathbf{H}$ vectors are shown for a plane-polarised monochromatic wave propagating in the z direction. The Poynting vector, $\mathbf{S}$ ($= \mathbf{E} \times \mathbf{H}$), which defines the energy flow is also shown. Although the radiation from an incandescent source has no preferred polarisation, the radiation from many astronomical sources is partially polarised either intrinsically, or by passage through the interstellar medium. In astronomy, the direction of the wave vibrations is taken to be that of the electric vector [90]. The measurement of polarisation is discussed in Chapters 3 and 7.

Within this century it has been recognised that a wide range of radiations from gamma rays to radio can be explained as electromagnetic radiation. They differ from each other only in wavelength. In 1900 Planck was able to show that the energy of electromagnetic waves must be quantised into discrete packets or 'photons' whose energy conforms to the relationship

$$E = hv \quad \text{J} \tag{1.1}$$

where h is a constant (6.6262×10^{-34} J s) and v is the frequency in hertz (cycles per second) of the associated vibrations of the electric and magnetic vectors.

Table 1.1 summarises the various recognised types of electromagnetic radiation, their range in wavelength and frequency, the range in energy of their associated photons, and the temperature which is characteristic of the photon energies ($T = hv/k$, where k is Boltzmann's constant).

One remarkable feature of Table 1.1 is the very limited spectral range of the optical region compared to the others. From a physiological point of view, this is understandable since it is the only spectral region of unattenuated sunlight. Wavelengths shorter than 325 nm can only be

Fig. 1.1 The relations between electric and magnetic vectors, **E** and **H**, for a simple harmonic plane-polarised wave propagating in the z direction. The energy flow is given by **S**, the Poynting vector, where $\mathbf{S} = \mathbf{E} \times \mathbf{H}$. In astronomy the direction of the wave vibrations is taken to be that of **E**. The wavelength, λ, is indicated.

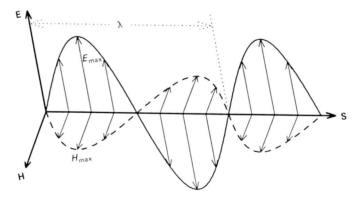

Table 1.1 *The electromagnetic spectrum*

Radiation	Wavelength	Frequency (Hz)	Energy/photon (J)	Temperature (K)
Gamma rays	<0.1 nm	$>3 \times 10^{18}$	$>2 \times 10^{-15}$	$>10^{8}$
X-rays	0.001 nm–100 nm	3×10^{20}–3×10^{15}	2×10^{-13}–2×10^{-18}	10^{10}–10^{5}
Ultraviolet	10 nm–300 nm	3×10^{16}–10^{15}	2×10^{-17}–7×10^{-19}	10^{6}–5×10^{4}
Optical	300 nm–1 μm	10^{15}–3×10^{14}	7×10^{-19}–2×10^{-19}	5×10^{4}–10^{4}
Infrared	1 μm–1 mm	3×10^{14}–3×10^{11}	2×10^{-19}–2×10^{-22}	10^{4}–10
Microwave	1 mm–3 cm	3×10^{11}–10^{10}	2×10^{-22}–7×10^{-24}	10–0.5
Radio	1 mm–30 m	3×10^{11}–10^{7}	2×10^{-22}–7×10^{-27}	10–5×10^{-4}

observed from rocket or satellite altitude while atmospheric water vapour makes infrared and microwave observations possible only from high, dry sites or from balloon or satellite altitude. The atmosphere is very transparent for radio observations at wavelengths shorter than 10 m.

Several astronomies

There is a tendency to subdivide observational astronomy into the areas of X-ray, optical, infrared, radio, etc. This is in part historical since the opening up of a new region of the spectrum for observation usually involves some major step in technology, for example the use of rockets and satellites for X-ray observations, or sensitive bolometers in the infrared. But it is clear from the characteristic temperatures given in the fifth column of Table 1.1 that the individual spectral regions also tend to be appropriate to the observation of quite different types of object and radiative processes.

Figure 1.2 shows four images to the same scale and orientation of M31, the Andromeda galaxy, from different spectral regions. Each region is sensitive to a different component of the galaxy.

(a) The direct, blue photograph shows the distribution of stars and brighter regions of ionised hydrogen (H II regions), with the stars being concentrated towards a nuclear bulge and to the spiral arms.

(b) The 21 cm image shows the distribution of neutral hydrogen (H I) which is entirely confined to the spiral arms [122].

(c) The 60 μm image was made by the Infrared Astronomical Satellite (IRAS is described in Chapter 3) and the radiation probably comes largely from warm interstellar dust grains at 30 to 40 K [170]. The properties of dust grains are discussed later in this chapter. Although also concentrated in the spiral arms there is an important component of the warm grains in the nuclear bulge. The 60 μm map of M31 closely resembles long wavelength radio maps made at, say, 50 cm [170]. The latter tend to show the distribution of non-thermal (synchrotron) radiation which arises from relativistic electrons confined by interstellar magnetic fields (see later in this chapter).

(d) This shows some of the 80 unresolved X-ray point sources detected by the Einstein observatory (described in Chapter 3) [396]. Most are probably compact objects such as supernova remnants (neutron stars or black holes) or white dwarf stars in binary systems with plasma temperatures of several million degrees [63]. Some of them vary in X-ray brightness and, while there is a strong concentration of them towards the nuclear bulge, some of those further from the nucleus can be identified with globular clusters and bright blue supergiant stars in star forming regions.

Astronomical observations can be divided into the two broad categories of survey and high resolution. The first instruments built for a

Fig. 1.2 Four images of the Andromeda Galaxy, M31, taken at different wavelengths, (a) blue, 430 nm, Lick Observatory photograph, (b) 21 cm, published with permission from [122], (c) 60 μm, published with permission from [170], (d) 0.5 to 4.5 keV [396].

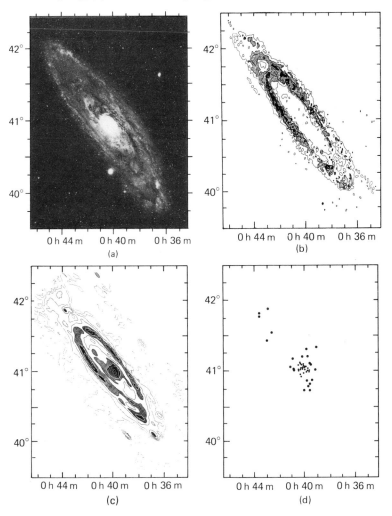

new spectral region usually carry out a survey to find bright or unusual sources. Subsequent instruments and programmes, particularly as techniques improve, examine these objects in detail. There are notable exceptions to this sequence. For example, in the rocket ultraviolet there has been no faint sky survey to support high resolution instruments such as the Hubble Space Telescope (HST is described in Chapter 3).

Thermal radiation from an ideal source

The energy emitted from unit area of a black body (defined as a surface which absorbs all radiation falling on it) at an absolute temperature T kelvin is equal to the amount of energy falling on one side of unit area inside a cavity also at temperature T [229]. The total energy is given by

$$F = \sigma T^4 \quad \mathrm{J\,s^{-1}\,m^{-2}} \tag{1.2}$$

where σ is the Stefan–Boltzmann constant ($5.670 \times 10^{-8}\,\mathrm{J\,m^{-2}\,K^{-4}\,s^{-1}}$) and T is known as the effective temperature. The spectral energy distribution, $F(\lambda)$ or $F(\nu)$, depends only on the temperature and is given by the Planck function

$$F(\lambda) = 2\pi hc^2 \lambda^{-5}/(e^{hc/\lambda kT} - 1) \quad \mathrm{J\,s^{-1}\,m^{-3}} \tag{1.3a}$$

$$F(\nu) = 2\pi h\nu^3 c^{-2}/(e^{h\nu/kT} - 1) \quad \mathrm{J\,s^{-1}\,m^{-2}\,Hz^{-1}} \tag{1.3b}$$

where $k = 1.3807 \times 10^{-23}\,\mathrm{J\,K^{-1}}$ and λ is in metres.

The Planck function defines both the shape and the intensity of the spectral energy distribution from unit area of a black-body source. Curves for the temperatures $10^4\,\mathrm{K}$ to $4 \times 10^3\,\mathrm{K}$ are shown at 1000 K intervals in Figure 1.3. The wavelength of maximum intensity moves towards shorter wavelength with increasing temperature according to the Wien displacement law

$$T\lambda_{\mathrm{max}} = \mathrm{constant} = 2.897 \times 10^{-3}\,\mathrm{m\,K} \quad (\lambda \text{ in m}) \tag{1.4}$$

Curves for different temperatures never intersect.

For wavelengths where $h\nu/kT \ll 1$, i.e. the long wavelength tail of the energy distribution, the Planck function can be approximated by the classical Rayleigh–Jeans distribution

$$F(\lambda) = 2\pi ck T\lambda^{-4} \quad \mathrm{J\,s^{-1}\,m^{-2}\,nm^{-1}} \tag{1.5a}$$

or,

$$F(\nu) = 2\pi c^{-2} kT\nu^2 \quad \mathrm{J\,s^{-1}\,m^{-2}\,Hz^{-1}} \tag{1.5b}$$

Where $h\nu/kT \gg 1$, i.e. the short wavelength side of the energy distribution, in the quantum limited region, the Planck function can be approximated by the Wien distribution

$$F(\lambda) = 2\pi hc^2 \lambda^{-5}/e^{hc/\lambda kT} \quad \text{J s}^{-1} \text{ m}^{-2} \text{ nm}^{-1} \tag{1.6a}$$

or,

$$F(v) = 2\pi hc^{-2} v^3/e^{hv/kT} \quad \text{J s}^{-1} \text{ m}^{-2} \text{ Hz}^{-1} \tag{1.6b}$$

Dilute radiation

The so-called 3 K cosmic background radiation (actually closer to 2.7 K) is the only cosmic source which approximates to the flux of radiation expected within a closed cavity. All other sources subtend solid angles which are much less than 4π steradians and the radiation received from them is said to be dilute.

If the irradiance or radiation received from a circular source which subtends an angle α radians is f (J m^{-2} s^{-1}) then its radiance or surface brightness is defined to be

$$B = f/(\pi \alpha^2/4) \quad \text{J m}^{-2} \text{s}^{-1} \text{sr}^{-1} \tag{1.7}$$

Fig. 1.3 The spectral energy distributions for black-body sources in the temperature range 4000 K to 10^4 K. The displacement of the maximum intensity to shorter wavelengths with increasing temperature according to the Wien law is also shown.

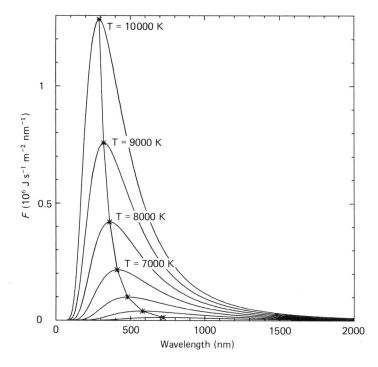

The equivalent black-body radiation for such a source can be derived from the geometry shown in Figure 1.4. The square represents a one-square metre plane surface and θ measures angles from the normal to the surface while ϕ is the angle of rotation about the normal. The radiation falling on the surface at an angle θ from an infinitesimal solid angle $d\omega$ from a source of surface brightness B is $B\cos\theta\sin\theta\,d\theta\,d\phi$, where $\cos\theta$ gives the projected area in the direction θ, and $d\omega = \sin\theta\,d\theta\,d\phi$. The total radiation, F, falling on the surface from a hemisphere of surface brightness B is simply

$$F = \iint B\sin\theta\cdot\cos\theta\cdot d\theta\cdot d\phi = \pi B \quad \mathrm{J\,s^{-1}\,m^{-2}} \tag{1.8}$$

Hence, from (1.7) and (1.8), if the source radiates as a black body the black-body radiation flux is given by

$$F = \pi B = 4f/\alpha^2 \quad \mathrm{J\,s^{-1}\,m^{-2}} \tag{1.9}$$

where α is in radians. The same expression holds true for the monochromatic flux

$$F_\lambda = 4f_\lambda/\alpha^2 \quad \mathrm{J\,s^{-1}\,m^{-2}\,nm^{-1}} \tag{1.10}$$

Stars

The spectral energy distribution in the optical and ultraviolet regions (138 to 1080 nm) is plotted for the bright northern star Vega in Figure 1.5 [209, 213, 307]. (The observations were made at wavelengths which are comparatively free from strong stellar absorption lines.) The technique for the ground based observations involves the direct comparison of a black-body source of known geometry with radiation

Fig. 1.4 Geometry used in the discussion of dilute radiation.

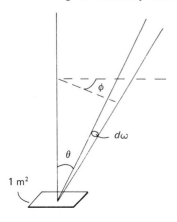

from a star [187, 220, 227, 255, 388, 419]. Figure 1.6 shows a section of one such source which used the freezing point of platinum as the standard [388]. Diaphragms between 0.5 and 2 mm diameter defined the angular size of the source which was placed 2.5 km from the telescope. With the telescope looking alternately at the star and the black body their spectral intensities were measured at the selected wavelengths. The vertical atmospheric extinction was measured by observing the star through different air masses in the manner described in more detail in Chapter 2.

The discontinuity between the satellite (UV) and ground based (optical) observations near 300 nm in Figure 1.5 probably reflects a UV calibration problem. The latter is less direct than the ground based and more subject to systematic error. On the other hand, the marked discontinuity at 380 nm is real.

The curvature of the spectrum between 400 nm and 800 nm conforms quite well to that expected from a black body at a temperature of 16 000 K

Fig. 1.5 The observed spectral energy distribution of Vega between 138 nm and 1080 nm from ground based (triangles) and satellite observatory (squares) observations. Black-body curves corresponding to 16 000 K and 10^4 K are shown with dilution factors of 2×10^{-17} and 5.28×10^{-17}, respectively.

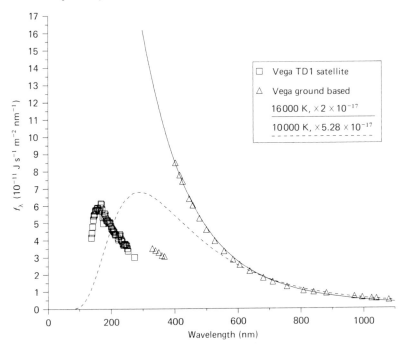

but differs from it in scale by a factor of 2×10^{-17}. For wavelengths >600 nm a black-body or Planck curve for $T = 10\,000$ K fits the observations best after it has been scaled by a factor of 5.3×10^{-17}. By substituting $f/F = 5.3 \times 10^{-17}$ for Vega in equation (1.3) we get a value of $\alpha = 3 \times 10^{-3}$ arcsec, where α is the angular diameter. This is exactly the value found independently by interferometric techniques (see Chapter 4).

If the scale factor of 2×10^{-17} for $T = 16\,000$ K is used instead we get a value of $\alpha = 1.8 \times 10^{-3}$ arcsec. The ambiguity in assigning one or the other of the two possible Planck curves arises because a star does not radiate as a black body with a unique surface temperature.

Energy generated at the centre of the star by nuclear fusion leaks to the surface in some 10^6 years following the temperature gradient. The degree of scattering or absorption in the final, outer, layers or atmosphere of the star depends upon the chemical composition and state of excitation of the atmosphere. Photons with energies appropriate to electronic transitions in the ions, atoms, and molecules (for cooler stars) making up the atmosphere are scattered more frequently than others and, as a result, have shorter mean free paths than those associated with the continuous part of the spectrum. Consequently, photons associated with electronic transitions originate closer to the surface of the star and are characteristic of a lower temperature. For a star like Vega with a surface temperature of 10^4 K the principal sources of opacity in the optical region are electron

Fig. 1.6 Section of a black-body source used in the direct calibration of the Vega spectral energy distribution. Published with permission from [388].

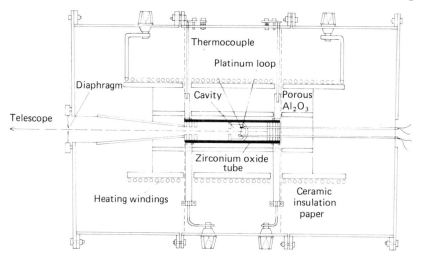

scattering and ionisation from the $n=3$ level of neutral hydrogen (Paschen continuum).

Crowding of the higher members of the Balmer series of neutral hydrogen (transitions from level $n=2$) together with the onset of ionisation for wavelengths <365 nm (Balmer continuum) leads to the large discontinuity in flux near 380 nm in Figure 1.5. This means that, on the short wavelength side of the discontinuity, radiation is being received from a higher, cooler, layer in Vega's atmosphere than on the long wavelength side.

The temperature determined by fitting a Planck curve to a portion of the observed spectrum is known as a *colour temperature*. When the entire spectrum is known it is possible to derive what is called an *effective temperature* [94], by using the Stefan–Boltzmann relation together with equation (1.9)

$$F = \pi B = 4f/\alpha^2 = T^4 \sigma \quad \text{J s}^{-1}\,\text{m}^{-2} \tag{1.11}$$

where $f = \int f_\lambda \,d\lambda$. Although Vega has not been detected at radio frequencies it is possible to make some estimate of f and, using the interferometrically measured value of α, this leads to $T_{\text{effective}} = 10^4$ K, on applying equation (1.11), which is consistent with the near infrared colour temperature derived above.

Alternatively, since fitting a Planck curve to the spectral energy distribution in the infrared appears to give a good estimate of effective temperature, one can derive an angular diameter from equation (1.9) as was done in the first paragraphs of this section [37]. The infrared spectral region is much less effected by strong absorption features and, being on the Rayleigh–Jeans tail of the black-body curve, the fitting of a black-body curve is much less sensitive to the temperature.

The continuous opacity sources for a star such as Vega (Balmer and Paschen continua and electron scattering) are well enough understood that there is good agreement between the emergent fluxes predicted from model atmosphere calculations and the observations. The results of such a model calculation are shown superimposed on the observations in Figure 1.7 [348].

Magnitudes

In the second century BC Hipparchus catalogued about 1000 of the visible stars. He classified them into six categories of brightness with those in the first category being the brightest. In 1856 Pogson proposed that these categories be called magnitudes and pointed out that what appear to be equal intervals in brightness to the eye are in fact equal ratios in light flux. Further, it turned out that a difference of five of Hipparchus' categories corresponded to a ratio of about 100:1 in light flux. The system of magnitudes was formalised to make the ratio exactly 100:1. Hence, by definition, if two stars give intensities I_1 and I_2 when observed with the same photometer, the difference in magnitude, Δm, is

$$\Delta m = m_1 - m_2 = -2.5 \log_{10}(I_1/I_2) \tag{1.12}$$

where $_1$ and $_2$ refer to the two stars. In other words a difference of one magnitude corresponds to a brightness ratio of 2.512.

Although the magnitude scale was developed only as a convenient system of visual photometry to reflect the logarithmic response of the eye, optical astronomers persist in expressing relative stellar brightness and colours in magnitudes. This has certain advantages for mental arithmetic. For example a factor of ten corresponds to 2.5 magnitudes and, where the two fluxes I_1 and I_2 are similar one can use the approximation

$$I_1/I_2 = 1 - \Delta I \tag{1.13}$$

Fig. 1.7 The observed spectral energy distribution of Vega (triangles) is compared with the emergent flux predicted from a model atmosphere calculation (solid line) [348].

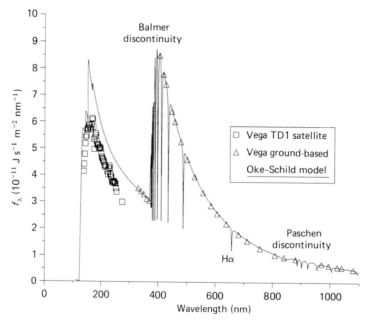

where $\Delta I \ll 1$, hence

$$\Delta m = 2.5(\log_{10}(1 + \Delta I))/(\ln 10) = 1.086\Delta I \qquad (1.14)$$

which means that when two stars are of similar brightness their magnitude difference is approximately equal to their fractional difference in brightness.

Magnitudes are also used to describe stellar energy distributions. The colour of a star is defined as

$$m(\lambda_2) - m(\lambda_1) = -2.5 \log(aI_1/bI_2) \qquad (1.15)$$

where λ_1 and λ_2 refer to the effective mean wavelengths at which the intensities I_1 and I_2 were measured and a and b are constants peculiar to the equipment.

Magnitude systems

Stellar magnitudes and colours are calibrated in terms of standard stars which have been established through photoelectric photometry and, in many cases, tied to the bright stars. As a rough guide, in the most widely used broad band photometric systems the magnitude of Vega is zero at all wavelengths at which it has been detected. This means that its colours based on these systems are also zero which can be misleading because zero colour does not correspond to a 'flat' spectral energy distribution.

Table 1.2 lists the bandpasses, and mean wavelengths, defined by the combined detector, filter, telescope, and atmospheric transmission, of some of the more commonly used photometric systems. Apart from the U filter whose bandpass is largely confined to the short wavelength side of the Balmer discontinuity, the choice of bandpass and mean wavelength for the broad band filters in Table 1.2 were largely instrumental and reflect the availability of detectors sensitive to the various spectral regions, and suitable filter materials. The B and V filters were chosen to match the peak of the normal photographic response and of the eye, respectively, when used in combination with an S-11 (see Chapter 7) photocathode. The infrared bandpasses are restricted by the windows in the atmospheric transmission as shown for J, H, and K (and the j, h, k, filters adopted at Kitt Peak) in Figure 1.8 [272]. Filters are discussed in more detail in Chapter 7.

Inverse square law

One of the most important elements of an astronomical source is its distance. Cosmic distances are very large when compared with the scale of the solar system. The star nearest to the Sun is Proxima Centauri (4.3 light years away) from which the radius of the Earth's orbit subtends a maximum angle (known as the annual parallax) of only 0.76 arcsec. For the much brighter star Vega at 26.5 light years the angle is 0.123 arcsec. The unit of distance adopted by astronomers is the parsec (pc) which is defined as the distance at which the semi-major-axis of the Earth's orbit

Table 1.2 *Photometric filter characteristics*

System	Filter	λ (nm)	$\Delta\lambda$ (nm)
UBV (Johnson–Morgan)	U	365	70
	B	440	100
	V	550	90
Infrared (Johnson)	R	700	220
	I	880	240
	J	1250	380
	K	2200	480
	L	3400	700
	M	5000	1200
	N	10400	5700
ubvyβ (Strömgren–Crawford)	u	350	34
	v	410	20
	b	470	16
	y	550	24
	β	486	3, 15

Fig. 1.8 The Kitt Peak summer atmospheric transmission in the near infrared. The transmission curves for the Johnson J and K filters, and the Kitt Peak J, H, and K, filters (indicated by j, h, and k) are shown by the dashed lines. (Reproduced from [272].)

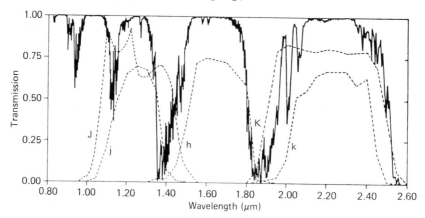

about the Sun (known as an Astronomical Unit $= 1.495\,98 \times 10^{11}$ m) subtends an angle of one arc second (i.e. 1 pc $= 206\,264.8$ AU $= 3.2615$ light years $= 3.0856 \times 10^{16}$ m). This makes the distances to Proxima Centauri and Vega 1.32 and 8.13 parsecs, respectively, or simply the inverse of their parallaxes in arc seconds.

The calculation of distances to more remote objects for which the parallax is too small to be measured accurately requires the assignment of an intrinsic or absolute luminosity to the source. It is identified with some equivalent local standard for which a distance is known. The distance is then estimated using the geometrical property that the irradiance of the source decreases as the inverse square of its distance from the observer, i.e.

$$(d/d_0)^2 = I_0/I \tag{1.16}$$

where d and I refer to the distance and irradiance of the remote source, respectively, and I_0 refers to the 'identical' standard at the known distance d_0.

The optical/infrared spectral regions are still the only ones in which there are recognisable standard candles, such as stars (variable and non-variable), H II regions, galaxies, and clusters of galaxies, which can be used to establish distances according to the inverse square law scheme outlined above.

Interstellar dust

The inverse square law is only applicable when radiative flux is conserved. This requires that the medium between the observer and the source be non-absorbing. Unfortunately, interstellar space is not quite empty. In our galaxy (as in all spiral galaxies) there is a tenuous interstellar gas (mostly hydrogen) with a density of about 8×10^6 atoms m^{-3} and fine, smoke-like, particulate matter with a density of about 10^{-23} kg m^{-3} or roughly 200 grains per cubic kilometre. The grains are very effective in attenuating optical and ultraviolet radiation. Consequently, before applying the inverse square law to the measured fluxes one must estimate the extinction by interstellar dust. The technique is discussed below.

Although direct starlight is reduced, starlight scattered by the dust grains contributes a detectable background from the Milky Way, while the absorbed radiation contributes to heating of the grains. Some dense, circumstellar dust clouds are bright sources in the infrared due to heating of the grains. In the visible and ultraviolet regions where the grains have a high albedo such clouds are often bright because of scattering of light from the central star, and they are known as reflection nebulae. Such scattering and re-emission often impose a polarisation on the scattered or re-radiated radiation.

The interstellar dust grains are probably elongated or needle shaped and they suffer some large scale alignment by interstellar magnetic fields. As a result transmitted starlight is also polarised, which means that the amount of extinction depends on the plane of polarisation and the polarisation is also a function of wavelength. The accurate measurement of polarisation is discussed in Chapter 7.

Absolute magnitudes and colour excess

It is possible to determine the photometric elements of most normal stars by measuring their magnitudes and a small number of colours. The absolute magnitude of a star, $M(\lambda)$, at a wavelength λ is defined as the magnitude which would be observed if the star were at a distance of ten parsecs from the observer. The observed or apparent magnitude, $m(\lambda)$, is related to the absolute magnitude by the inverse square law after making an appropriate correction for interstellar extinction

$$m(\lambda) - M(\lambda) = 5 \log_{10} d - 5 + A(\lambda) \tag{1.17}$$

where d is the distance to the star in parsecs, and $A(\lambda)$ is the total extinction by interstellar dust in magnitudes at the wavelength λ. The quantity $(m(\lambda) - M(\lambda))$ is known as the distance modulus.

In the absence of interstellar reddening the colours of normal, unevolved stars are a function of their surface temperature, chemical composition, and radius only. The variation of extinction with wavelength is remarkably similar in all parts of the Milky Way apart from certain H II regions and reflection nebulae. Colour excess, $E(\lambda_2 - \lambda_1)$, is defined as

$$E(\lambda_2 - \lambda_1) = (m(\lambda_2) - m(\lambda_1)) - (m(\lambda_2) - m(\lambda_1))_0 \tag{1.18}$$

where $(m(\lambda_2) - m(\lambda_1))_0$ refers to the unreddened or intrinsic colour and $(m(\lambda_2) - m(\lambda_1))$ is the observed colour.

Differential interstellar extinction curves are determined by comparing the colours of interstellar reddened stars with those of identical but unreddened stars. Normalised colour excesses, Δm, are plotted against λ^{-1} (since Δm varies linearly with λ^{-1} in the optical region), where

$$\Delta m = E(\lambda - V)/E(B - V) \tag{1.19}$$

Such a plot is shown in Figure 1.9 [274]. The total extinction, $A(\lambda)$, is calculated from the ratio of total to selective extinction, R, where

$$R = A(V)/E(B-V) \qquad (1.20)$$

and equation (1.17) can be rewritten as

$$m(\lambda) - M(\lambda) = 5 \log_{10} d - 5 + RE(B-V) \qquad (1.21)$$

R is one of the least well known 'constants' in astronomy. The currently accepted standard value is 3.2 ± 0.2 [290]. In theory R can be calibrated by extending measurements of the extinction curve to sufficiently long wavelengths where interstellar extinction can be considered negligible (the λ^{-4} Rayleigh 'tail'). Unfortunately such an extrapolation assumes that the reddened and unreddened stars have identical infrared spectra which is often not true. The presence of a red, optically faint companion, heating of circumstellar dust, and stellar atmospheric effects can all lead to an infrared excess.

Alternatively, when distance modulus is plotted against colour excess for individual stars in a cluster (i.e. all at the same distance) the gradient should be R [406]. This technique requires a good absolute magnitude calibration and a wide range of colour excess among the cluster members and the cluster must be sufficiently distant that a correlation of extinction

Fig. 1.9 The interstellar reddening curve. Normalised colour excess is plotted as a function of λ^{-1} for the optical, UV, and near infrared spectral regions. The curves for two stars with unusual reddening are also shown. The ultraviolet observations were made with the satellite OAO-2, and $\times$ with Copernicus. (Published with permission from [274].)

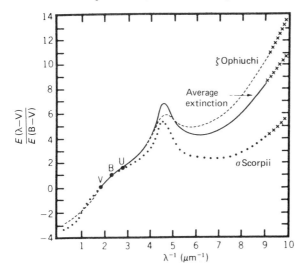

with depth in the cluster does not introduce a significant, systematic, error.

Figure 1.10 shows the two-colour plot of $(U - B)$ against $(B - V)$ for some 46 000 stars [304]. The majority of them lie in an S-shaped

Fig. 1.10 The two-colour plot of $(U - B)$ against $(B - V)$ for 46 000 stars. (Published with permission from [304].)

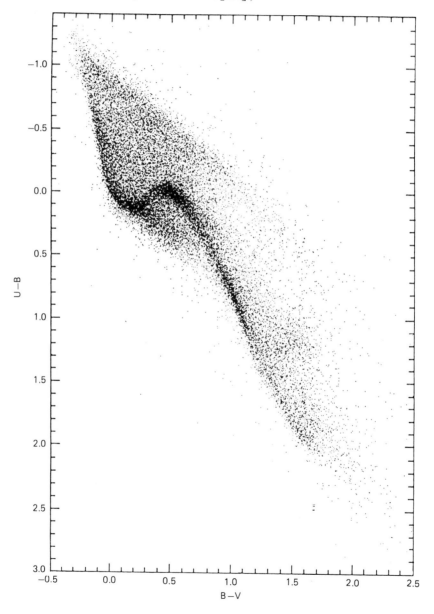

distribution, the main sequence. They are unevolved, unreddened by interstellar dust, and have a similar metal content to the Sun. Points for intrinsically blue stars (spectral types O and B) lie towards the top left corner with the gradient of the reddening line shown by the diagonal upper envelope to all the points. Increasing interstellar reddening displaces points along lines of gradient $E(U-B)/E(B-V)$ ($=0.72$ to 0.8 depending on the intrinsic colour of the star).

The locus of points for black bodies of different temperature would follow a diagonal line similar to the reddening locus. The S-shape of the main sequence is caused mostly by absorption in the Balmer continuum (see Figure 1.7). Population II stars have a lower metal abundance than the younger Population I stars such as the Sun. The blue/UV atmospheric opacity is significantly less for Population II than for Population I among the cooler stars. Points for the former lie above the sequence defined by the latter in Figure 1.10.

Bolometric magnitudes and bolometric corrections

The absolute magnitude normally quoted is M_V. Since it is essentially monochromatic it is not a measure of the total radiation emitted by the star over the full spectrum [94]. The bolometric magnitude, B(bol), is the absolute magnitude which would be measured by an ideal bolometer exposed to all of the radiation from the star and having equal sensitivity at all wavelengths. B(bol) is calculated from the absolute magnitude by adding the bolometric correction, B.C., where

$$M(\text{bol}) = M_V + \text{B.C.} \tag{1.22}$$

by convention B.C. $= -0.07$ for a solar type star. In order to estimate B.C. one must know the emitted flux over the whole spectrum. For most stars this is largely possible by combining calibrated ultraviolet, visible, and infrared observations and inferring the residual radiation outside this range using model atmosphere predictions.

Figure 1.11 shows a plot of B.C. against temperature [169]. As might be expected, B.C. $= -0.07$ close to $T = 7000$ K is the maximum value of the bolometric correction because there is more radiation outside the V-band radiation for stars of other temperatures. Stars of higher temperature have more ultraviolet, while those of lower temperature have greater infrared.

Surface photometry

There is considerable astrophysical interest in relating the surface colours and brightness distributions of extended objects such as nebulae, clusters, and galaxies to those of stellar standards. Optical brightness distributions of extended sources are often calibrated in units of magnitudes per square arc second, m'', or magnitudes per square arc minute, m', where $m' = m'' - 8.89$ ($= 2.5 \log_{10} 3600$). If the brightness on a particular magnitude system is, say, 21 magnitudes per square arc second, this means that the brightness is equivalent to that of a 21 mag star spread over each square second of arc of the source.

Integrated brightness measurements are also sometimes quoted as the number of stars of a given magnitude per square degree which would have the same integrated brightness [5]. The integrated starlight in V at the north galactic pole, for example, is equivalent to some 30 tenth magnitude stars per square degree. In the plane of the Milky Way the figure is closer to 400.

The estimation of absolute magnitude

The spectra of stars were initially classified according to their appearance and assigned to a lettered category, each of which had ten subclasses. It was later recognised that these categories could be arranged according to temperature in the order O B A F G K M, with class O containing the hottest stars with temperatures $> 20\,000$ K, and with the

Fig. 1.11 The variation of bolometric correction with temperature for main sequence stars [169].

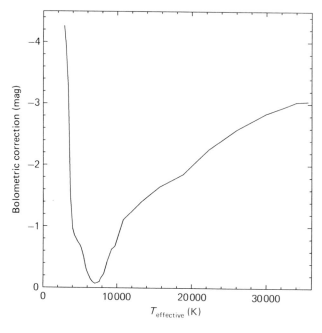

coolest stars being of type M (<3000 K). Vega is of type A0 ($10\,000$ K), while the Sun is G2 (6000 K), and Betelgeuse is M2 (3000 K). The variation of T_{eff} with spectral type is shown in Figure 1.12.

It was also recognised that at each spectral type there was a wide range in luminosity between the individual stars. A luminosity was assigned by a Roman numeral from VI to I with the majority of stars being of class V on the main sequence, class I being supergiants. Vega and the Sun are luminosity class V, and Betelgeuse class I. This system of classification is now known as the MK system after Morgan and Keenan who established its foundations [149]. Some representative spectra are shown in Figure 1.13.

Certain spectral features are sensitive to luminosity and, by carefully measuring their absorption strength, it is possible to calibrate and assign absolute magnitudes to individual stars. For example, the absorption strengths of individual members of the Balmer series of hydrogen in stars of spectral types O, B, and A are sensitive to the electron pressure and the degree of ionisation in the stellar atmosphere. The lines become broader and stronger with increasing surface gravity, i.e. lower luminosity. Therefore, the more negative the absolute magnitude the weaker the Balmer lines [295].

Fig. 1.12 Effective temperature, T_{eff} vs. spectral type.

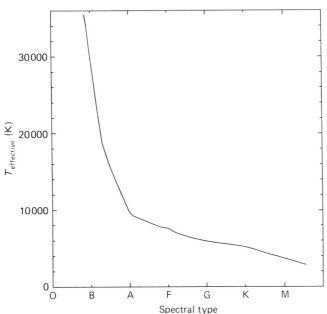

Nebulae and high temperature plasmas

Thermal radiation

Although the visible atmosphere of a star contains only some 10^{-11} of its mass, there are many extended sources which are optically thin over a wide range of wavelengths. The thermal radiation from a plasma is often termed thermal bremsstrahlung (literally braking radiation) and the emission coefficient, $j(v)$, is given by

$$j(v) = 5.44 \times 10^{-40}(g_{ff}n(e)n(p)\exp(-hv/kT))/T^{0.5} \qquad (1.23)$$

where the units are $\mathrm{J\ m^{-3}\ s^{-1}\ sr^{-1}\ Hz^{-1}}$. The Gaunt factor, g_{ff}, for

Fig. 1.13 Six optical spectra of different spectral types.

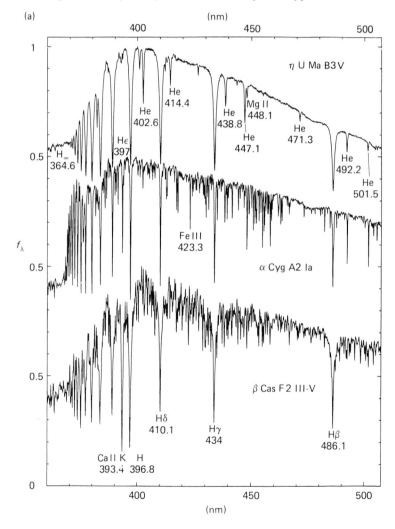

free–free transitions changes only slowly with v, $n(e)$ and $n(p)$ are the number of electrons and the number of protons per cubic metre, respectively. For an electrically neutral plasma $n(e) = n(p)$ and the product $n(e)n(p)$ can be replaced by $(n(e))^2$. From this it can be seen that the emission varies as the square of the electron density but only as the square root of the temperature.

The absorption coefficient, $k(v)$, is related to the emission coefficient through Kirchhoff's law such that

$$k(v) = j(v)/B(T) \quad \text{m}^{-1} \tag{1.24}$$

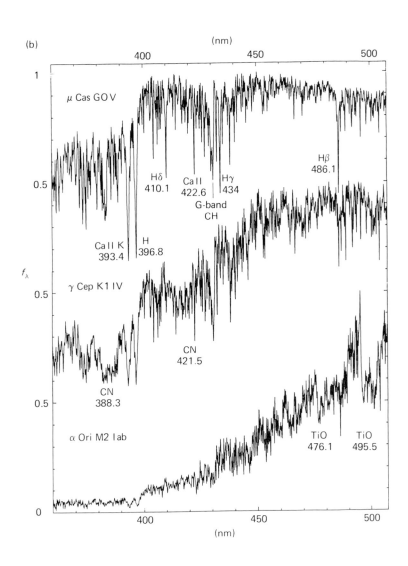

If v is small compared with kT/h then the Rayleigh–Jeans approximation can be used for $B(T)$ and, including the expression for g_{ff}, gives for $k(v)$ [366]

$$k(v) = 17.3(1 + 0.130 \log_{10}(T^{1.5}/v))n(e)n(p)T^{-1.5}v^{-2} \quad m^{-1} \quad (1.25)$$

which indicates that $k(v)$, and hence self-absorption, increases as v^{-2}.

In Figure 1.14 the radio spectrum of the Orion Nebula is shown [287]. The nebula is a low density plasma ($n(p) = n(e) = 6 \times 10^8$ m^{-3}), normally called an H II region, which is being heated by ionising radiation from the hot OB-type stars (temperatures $> 2 \times 10^4$ K) in the Trapezium (a close quartet of stars which make up the middle star in Orion's Sword). The curve at longer wavelengths has the characteristic shape of an optically thick source at some 10^4 K but the observations depart from the curve at a wavelength of about 0.5 m and remain roughly constant towards shorter wavelengths. This indicates that at wavelengths < 0.5 m the plasma is optically thin.

The only H II regions which are readily detected at radio frequencies subtend quite large solid angles. The Orion Nebula has almost the same area as the Sun (0.22 sq.degrees). There is, however, a class of compact, high temperature (10^6 K) plasmas which are detected over a wide spectral range from X-ray and gamma ray through radio. They are associated with the rapid heating of material in a stellar flare or by accretion onto a highly

Fig. 1.14 The radio spectrum of the Orion Nebula. (Published with permission from [287].)

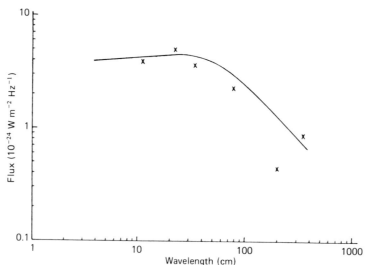

compact object in a binary system. Such sources often vary irregularly in brightness [33, 63, 82].

Emission lines

In addition to continuum radiation, continual recombination of the various constituent ions, together with collisional excitation in the plasma, leads to fluorescence of a wide range of emission lines. An optical spectrum of the Orion Nebula is shown in Figure 1.15. This shows and identifies some of the stronger emission lines. Emission lines are detected throughout the electromagnetic spectrum. For a detailed list see, for example, [246].

Densities are so low in interstellar space (10^{-21} kg m^{-3}) and in many H II regions, that lines which correspond to transitions normally considered forbidden in the laboratory, such as electric quadrupole transitions, are possible because ions can accumulate in metastable levels without being collisionally de-excited. Forbidden lines from electric quadrupole, magnetic dipole, and magnetic quadrupole transitions are designated by a square bracket enclosing the symbol for the ion, thus [O III].

A single square bracket is used for electric dipole intercombination lines, thus N II].

Fig. 1.15 An optical spectrum of the Orion Nebula with some of the principal emission lines identified.

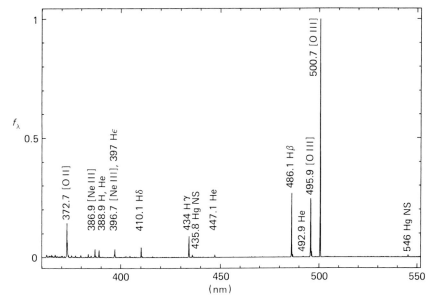

Non-thermal radiation
Synchrotron radiation [246, 366]

In astronomy the term 'non-thermal radiation' usually implies synchrotron radiation (or magnetobremsstrahlung) from electrons moving normally to a magnetic field with relativistic velocities (electron kinetic energies of the order of, or greater than the electron mass of 0.5 MeV). The radiation is highly directional being emitted in a small solid angle along the direction of motion of each electron and it is polarised with the electric vector in the plane of the electron orbit. The radiated spectrum depends upon:

(a) the electron energy distribution,
(b) the optical depth or degree of synchrotron self-absorption,
(c) the average magnetic field strength, and the 'age' of the synchrotron source (the time since the last injection of relativistic particles).

The radio spectra of most galaxies and quasars have the form

$$F(v) = v^{-\alpha} \tag{1.26}$$

where α is constant over a wide frequency range and is known as the spectral index. It can be shown that if the relativistic electrons have an exponential energy distribution of the form

$$n(E)dE = \text{constant} \cdot E^{-\gamma} dE \tag{1.27}$$

then

$$\alpha = (\gamma - 1)/2 \tag{1.28}$$

The formula for the high turnover frequency, v_2, is of the form

$$v_2 = \text{constant} \cdot H^{-3} \Delta t^{-2} \quad \text{Hz} \tag{1.29}$$

where H is the component of the magnetic field normal to the electron trajectories and Δt is the time which has elapsed since electrons of sufficiently high energy to emit at frequencies greater than v were ejected from the source. In this way the 'age' of a non-thermal source can be characterised by the high-frequency cutoff in its spectrum.

In Figure 1.16 the spectra of several bright non-thermal sources are shown together with those for several thermal sources. The non-thermal character is very clear with the radiated energy increasing towards long wavelengths until synchrotron self-absorption sets in. In the region of self-absorption the spectrum resembles a black-body source.

Synchrotron radiation is characteristic of some of the most energetic sources, particularly those of galactic dimensions or those associated with supernova remnants. As a result, radio observations have been very effective in detecting and mapping relativistic plasmas of the sort seen in association with active galaxies where nothing can be detected optically.

When electrons move parallel to curved magnetic field lines, so-called 'curvature-radiation' may be generated if the fields are both sufficiently large and strongly curved. Such conditions appear to exist in the vicinity of pulsars $(B \sim 10^8 \, \text{T}, \ R \sim 10^4 \, \text{m})$. Further discussion is contained in [373].

Inverse Compton scattering

When photons are scattered by electrons at rest the phenomenon is known as Compton scattering. Inverse Compton scattering (sometimes

Fig. 1.16 Spectra of several bright, non-thermal and thermal, radio sources. (Published with permission from [86].)

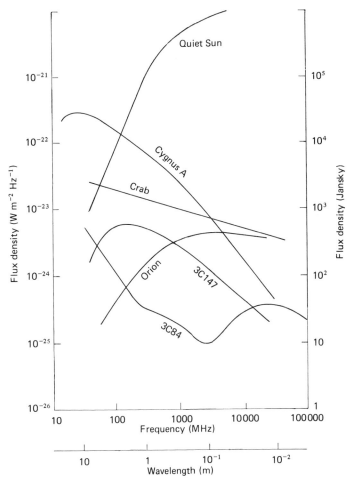
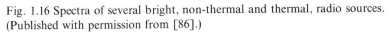

called electron scattering) involves the transfer of energy from highly energetic electrons to photons of much lower energy. When the electrons are relativistic with energies in the GeV range the photons may be boosted to gamma-ray energies. If the electron energy is E it loses energy at a rate given by the expression

$$dE/dt = \tfrac{4}{3}\sigma(T)c\rho(E/mc^2)^2 \quad W \tag{1.30}$$

where $\sigma(T)$ is the Thomson cross-section

$$\sigma(T) = \tfrac{8}{3}\pi r(e)^2 \quad m^2 \tag{1.31}$$

and $r(e) = e^2/mc^2$, the classical electron radius, m is the electron mass and ρ is the energy density of the low energy photons.

Although the total number of photons does not change, their spectral energy distribution is altered and the process favours energy loss by the most energetic electrons.

The 2.7 K cosmic background radiation provides a universal energy density $\rho = 4.8 \times 10^{-14}$ J m^{-3} (some 10^9 photons m^{-3}). This means that an electron with an energy in the GeV range would transform much of this energy into gamma or X-rays over intergalactic distances.

Čerenkov radiation

If a charged particle enters a medium in which its velocity exceeds the phase velocity of light in the medium, it will emit radiation into a cone of semi-angle θ such that [217]

$$\cos\theta = (\beta n)^{-1} \tag{1.32}$$

where $\beta = (1 - (v/c)^2)$, n is the refractive index, and v is the particle velocity.

Provided n varies slowly with λ the number of photons, N, emitted between the wavelengths λ_1 and λ_2 can be predicted quite accurately by a simple classical treatment [217]

$$N = 2\pi\alpha l (1/\lambda_1 - 1/\lambda_2)(1 - (\beta^2 n^2)^{-1}) \tag{1.33}$$

where α is the fine structure constant $(= 1/137)$ and l is the path length in the medium. Equation (1.33) implies that equal numbers of photons are emitted in each frequency interval, or, since the total energy emitted $= Nh\nu$, the number of photons emitted per unit path length increases linearly with frequency (which accounts for the blue colour of Čerenkov radiation). For example, about 15 photons would be emitted between the wavelengths 400 nm and 600 nm by a 500 keV electron moving through a 1 mm thickness of glass.

A significant number of cosmic ray primary and secondary electrons have sufficient energy to produce bursts of Čerenkov radiation in the

Earth's atmosphere and in refractive optics. Attempts have been made to detect high energy cosmic gamma-ray photons from the Čerenkov radiation emitted in the atmosphere by secondary particles [130].

For optical observations, the Čerenkov radiation produced by cosmic rays in the optics represents an unwanted background. The index of refraction for most refractive materials used in optics varies from unity in the near infrared to more than 1.6 in the ultraviolet where the transmission usually drops sharply. Since N increases with both n^2 and $1/\lambda$ the effect of the Čerenkov background increases rapidly into the UV. In order to minimise this background UV wavelengths not being measured (e.g. $\lambda < 320$ nm for ground based telescopes) should be filtered out when a detector which is sensitive to the UV is being used. Photon counting systems can also be gated by charged particle sensors.

The Zeeman effect

Emission and absorption lines arising in stellar atmospheres or the interstellar medium are distorted by the Zeeman effect or inverse Zeeman effect in the presence of a magnetic field [13, 14, 127, 146, 397]. In a weak, uniform field H (measured in tesla), the wavelength λ is shifted to $\lambda + \Delta\lambda$, where

$$\Delta\lambda = 4.67 \times 10^{-8} z H \lambda^2 \quad \text{nm} \tag{1.34}$$

z is calculated from the term values of the transition to take account of the effects of spin-orbit coupling on the classical Zeeman pattern. Values of z can range from 0 to 3 with the classical value being 1.

When the field is transverse to the observer's line of sight the classical pattern consists of a triplet (σ components) which has two linear components separated by $2\Delta\lambda$ with an undisplaced component in the orthogonal polarisation equal to the combined strength of the displaced components. Viewed parallel to the field the classical (π) pattern consists of only two components of equal intensity but oppositely directed circular polarisation and separated by $2\Delta\lambda$. The patterns, except for simple atoms, are normally more complicated than the classical [14]. The sign conventions and description of polarisation in terms of Stokes' parameters are given in Chapter 7.

Integrated fields for stars such as the Sun are typically a millitesla. Zeeman patterns are resolved for only a few stars which have both fields of a few tenths of a tesla and sharp metallic absorption lines [13]. Normally the patterns are unresolved and the fields are detected by measuring the displacements of the line between orthogonal linear polarisations or oppositely directed circular polarisations [127].

The λ^2 dependence of $\Delta\lambda$ means that much smaller displacements can be detected at radio wavelengths. This is fortunate as interstellar fields are typically of the order of a nanotesla [397].

2

Observational limits

Coherence [45, 178, 391, 433]

The example of Vega used in the discussion of stellar radiation in Chapter 1 emphasises the very small angular size of most cosmic sources and the fine structure expected within more extended sources. Telescopic images suffer from the effects of atmospheric perturbations in the optical and infrared spectral regions and from ionospheric distortions in the radio region. These distortions coupled with diffraction or other limitations in the telescope and detector make it impossible to resolve directly any stellar disc or the finer structure in radio sources. Nonetheless, the small true angular extent of these sources, coupled with the linear dimensions of the telescope, means that radiation from unresolved sources will be at least partially coherent.

Partially coherent radiation is subject to interference effects when superimposed on itself, either after suitable division using separate telescopes, or by division of the beam within a single telescope. In simple terms, if the angular size subtended by the source is smaller than the angular size of the first telescope diffraction minimum then the radiation will be partly coherent and some interference effects can be expected. (The term telescope is meant to cover any radiation collecting system and includes radio antennae.) The degree of coherence is a function of the ratio of the angular sizes of the source and the angular radius, θ, of the first minimum in the telescope diffraction pattern (the Rayleigh criterion)

$$\theta = 1.22\lambda/d \text{ radians} = 251\,643(\lambda/d) \text{ arcsec} \tag{2.1}$$

where d is the diameter of the telescope aperture (assumed to be circular). The actual distribution of light in the diffraction pattern is discussed in more detail in the section on apodisation in Chapter 3.

Single radio telescopes are only tens or hundreds of wavelengths across, while large optical telescope mirrors are measured in millions of wavelengths. It is particularly useful then, in the case of radio waves, to exploit the coherence property to improve angular resolution. The property is more difficult to exploit at optical wavelengths because of the scale of the phase distortion introduced by the Earth's atmosphere. For a 4 m telescope, at 500 nm, $1.22\lambda/d = 0.03$ arcsec. In this case the radiation from Vega ($\alpha = 0.003$ arcsec) would be highly coherent. In order to begin to resolve the disc of the star in the absence of the Earth's atmosphere d would have to be > 10 m.

Figure 2.1 shows the classical two-beam interference experiment in which the monochromatic image of a small, distant source (such as a star) is focussed onto a screen, S, by the lens, L, illuminated by the two apertures M_1 and M_2. The apertures are separated by a distance d and the source is in the direction θ as shown. If the source is completely coherent the intensity distribution on the screen consists of a series of interference fringes whose intensity varies as $(1 + \cos\phi)$, where ϕ, the difference in phase introduced by the geometrical difference in path length $d\sin\theta$, is given by

$$\phi = (2\pi/\lambda)d\sin\theta \tag{2.2}$$

In monochromatic light of wavelength λ the fringe maxima would occur when

$$d\sin\theta = n\lambda \tag{2.3}$$

where n is zero or an integer. I_{max} is measured at the centre of symmetry of the fringe pattern ($n=0$), while I_{min} is measured half a fringe cycle away ($n=\frac{1}{2}$). Fringe visibility, $V(d)$, is defined as

$$V(d) = (I_{max} - I_{min})/(I_{max} + I_{min}) \tag{2.4}$$

Fig. 2.1 A classical two-beam interference experiment in which a monochromatic image of a small, distant source is focussed onto the screen, S, by the lens, L, through two apertures M_1 and M_2 separated by a distance d. The source is in the direction θ.

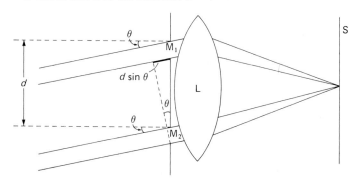

and is equivalent to the degree of coherence [45, 175]. The variation of $V(d)$ with d is shown in Figure 2.2 for a uniform disc of angular diameter θ. The fringes first disappear (i.e. $V(d)=0$) when

$$d = 1.22\lambda/\theta \tag{2.5}$$

which is identical to the Rayleigh criterion (equation (2.1)) for a telescope of diameter d. Since equation (2.5) is for a uniform disc it implies that the two apertures give twice the resolution (in the direction of d) as a telescope of the same diameter.

Figure 2.3 shows monochromatic fringe patterns obtained from a circular source with a two-beam interferometer in which the separation, d, has been varied [45, 379a]. The degree of coherence is indicated under each interferogram and it illustrates the progression through positive, zero, and negative fringe visibility as coherence is decreased.

The envelope of the fringe pattern is defined by the single aperture image of the source formed by M_1 or M_2 (assuming that both apertures have the same diameter). In the limiting case of coherent, nearly monochromatic radiation the number of fringe maxima within the central diffraction maximum, n, is given approximately by:

$$n = 2(1 + d/m) \tag{2.6}$$

where m is the diameter of M_1 or M_2.

Fig. 2.2 The variation of the fringe visibility, $V(d)$, with d for a uniform disc of angular diameter θ. (Published with permission from [175]).

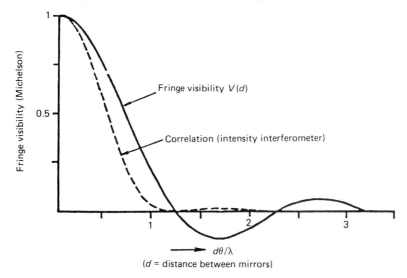

Fringe visibility (Michelson)

Fringe visibility $V(d)$

Correlation (intensity interferometer)

$d\theta/\lambda$

(d = distance between mirrors)

Coherence time and length

Only the zero-order fringes $(n=0)$ in equation (2.3) are in phase for all wavelengths contained within the bandpass $\Delta\nu$. As the difference in optical path length increases, the phases of the fringe patterns for different wavelengths become increasingly random. This reduces fringe visibility and produces a 'white' fringe pattern. The envelope of the 'white' fringe pattern is simply the Fourier transform of the bandpass $\Delta\nu$.

Fig. 2.3 Monochromatic fringe patterns obtained with an arrangement similar to that shown in Figure 2.1. The degree of coherence has been varied by changing d. (Published with permission from [379a].)

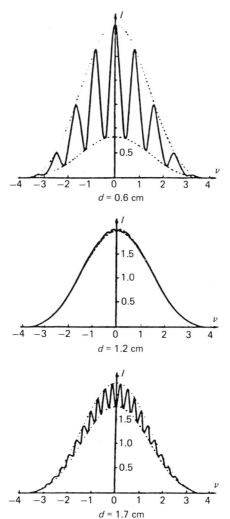

As $d \sin \theta$ (see Figure 2.1) increases, the fringe visibility is reduced by the sync function $g(\Delta v, t)$

$$g(\Delta v, t) = \sin(\pi \Delta v \cdot t)/(\pi \Delta v \cdot t) \tag{2.7}$$

where $t = (d \sin \theta)/c$, the time delay between the two beams, and Δv is the bandpass of the radiation. The actual fringe visibility is given by $V(d)g(\Delta v, t)$.

The factor $g(\Delta v, t)$ becomes zero when

$$t = \Delta t = 1/\Delta v \tag{2.8}$$

where Δt is known as the coherence time and the related coherence length Δl is given by

$$\Delta l = c/\Delta v \tag{2.9}$$

The coherence length, at optical wavelengths, is a few micrometres for a bandpass of a few tens of nanometres, and is the limiting difference in path length, $d \sin \theta$, if fringes are to be seen.

Photon flux

Information about any spectral energy distribution is both conveyed and, at shorter wavelengths, limited, by the number of photons available for observation. The photon distributions with wavelength, $N(\lambda)$, or frequency, $N(v)$, are simply $F(v)/hv$ or $\lambda F(v)/hc$, where

$$N(\lambda) = 2\pi c \lambda^{-4}/(\exp(hc/\lambda kT) - 1) \quad \text{photons s}^{-1}\,\text{m}^{-3} \tag{2.10a}$$

or,

$$N(v) = 2\pi v^2 c^{-2}/(\exp(hv/kT) - 1) \quad \text{photons s}^{-1}\,\text{m}^{-2}\,\text{Hz}^{-1} \tag{2.10b}$$

where wavelength is in metres.

The photon spectrum of Vega is plotted against wavelength as photons $\text{s}^{-1}\,\text{m}^{-2}\,\text{nm}^{-1}$ in Figure 2.4. This is a more useful plot than f_λ against λ (Figure 1.5) when discussing instrumental performance. Also, in Figure 2.5 n_λ is plotted for two black-body temperatures of 2000 K and 10^7 K. In both cases the source is assumed to have the same irradiance as Vega, which means that the ratio of the squares of their angular diameters must be $f(2000)/f(10^7) = (1000/10^7)^4 = 1.6 \times 10^{-15}$ by application of the Stefan–Boltzmann relation (equation (1.2)). The photon flux at maximum is the same although at quite different wavelengths. This follows directly from an application of Wien's displacement law and the Planck function.

Photon noise

The random fluctuation in the photon flux emitted by a steady thermal source can be described by a Poisson distribution. If the

Fig. 2.4 The photon spectrum, N_λ, of Vega. This is a more useful plot than Figure 1.5 for the discussion of instrumental performance.

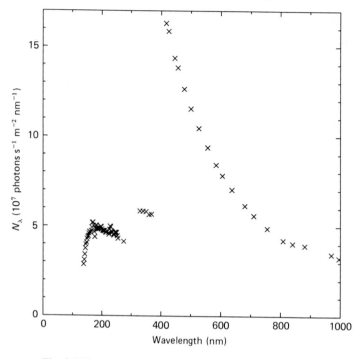

Fig. 2.5 The photon spectra, N_λ, for two sources one of 2000 K the other of 10^7 K. Each has the same irradiance as Vega.

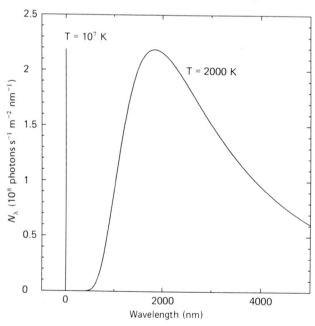

probability of *n* photons from the source crossing a given area in *t* seconds is $p(n, t)$ then

$$p(n, t) = (Nt)^n e^{-Nt}/n! \tag{2.11}$$

where N photons s^{-1} is the average flux from the source across the same area. Figure 2.6 shows the Poisson distributions for $(Nt) = 1$, 4, and 10.

This distribution has the important property that the root-mean-square (r.m.s.) fluctuation or 'shot noise' in the average flux N is simply

$$\text{r.m.s. noise,} \quad \sigma = N^{0.5} \tag{2.12}$$

It is clear from the unequal areas under the curves for the sources of equal irradiance at 2000 K and 10^7 K in Figure 2.5 that for sources which emit the majority of their radiation in the X-ray and gamma-ray regions, photon shot noise will dominate the detection process.

The photon noise limitation in the short wavelength regions is accentuated by the technical dependence of the telescope collecting area on wavelength. The figuring tolerance for the reflecting surfaces is directly proportional to wavelength. For example, the figuring tolerances for a 100 m radio dish and a 6 m optical mirror are roughly in the ratio $1 : 5 \times 10^{-4}$, while their apertures are in the ratio $1 : 6 \times 10^{-2}$. This means that, despite its smaller aperture, the figuring tolerance for the optical telescope is, relatively, more severe in terms of the aperture size. The largest collecting area available in a focussing telescope for X-rays is 0.04 m^2 (Einstein Observatory described in Chapter 3), compared with, say, the 100 m radio telescope with an area of some 1800 m^2, i.e. a ratio of $1 : 200\,000$ in collecting area and some 10^{15} RF photons/X-ray photon (depending on λ) for a given energy detected.

Wave noise

For bandpass limited, partially coherent radiation, a wave noise contribution must be included with the photon noise. Photons tend to bunch or cluster and are subject to Bose–Einstein statistics. When the bandpass or spectral energy distribution is Gaussian, the probability of detecting a second photon (of the same polarisation) is double that predicted by a Poisson distribution during the $0.2\Delta t$ following the initial detection as shown in Figure 2.7 (where Δt is the coherence time of equation (2.8)) [271]. In a classical picture the effect is known as wave noise and is caused by beating between neighbouring frequencies included within the band width of the receiver. Wave noise is discussed more fully in Chapter 6 in the description of the intensity interferometer and in

Fig. 2.6 Poisson distributions for $Nt=$(a) 1, (b) 4, and (c) 10, where p is the probability that a particular number of photons will be detected.

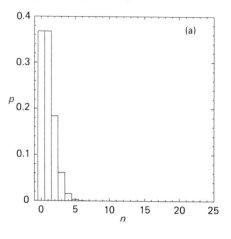

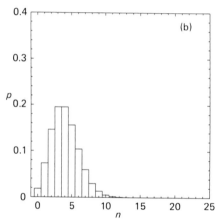

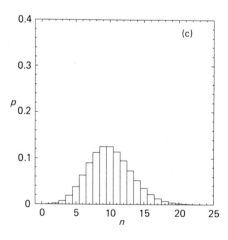

Chapter 7 in the discussion of radio reception. *The important distinction between wave noise and shot noise is that the former is proportional to N while the latter is proportional to $N^{0.5}$.*

The ratio of wave noise, $\sigma(w)$, to shot noise, $\sigma(p)$, in the output signal of a detector, for a given polarisation and from a coherent source, is [175, 176]

$$\sigma(w)^2/\sigma(p)^2 = n \qquad (2.13)$$

where n is the total number of photons actually detected per second per hertz.

In astronomy the contribution of wave noise is not significant at optical wavelengths because of the low photon fluxes normally encountered. For example, at 500 nm for a flux of 10^6 photons per second and a bandpass of 10 nm (10^{13} Hz) the contribution of wave noise to the photon shot noise would be of the order of 1 part in 10^7 if the time resolution of the detector is 10^{-9} s (1 ns) [175].

On the other hand wave noise greatly exceeds photon shot noise at radio frequencies where the number of photons necessary for a detection is many orders of magnitude larger and contained within a narrower bandpass [61]. The statistical limit to the precision of radio wave reception is discussed in Chapter 7 on receivers.

Fig. 2.7 The conditional probability of detecting a second photon of the same polarisation as a function of time after detection of the first photon in a plane polarised wave with a Gaussian spectral profile of width Δv. The probability is normalised by the probability predicted from a Poisson distribution. Adapted from [271].

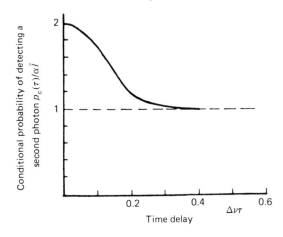

Detection

There are two ways in which radiation can be detected. The choice depends largely on the energy of the incident photons.

Incoherent detection

Sufficiently energetic photons can be detected incoherently either by the photoelectric effect, in which individual photons excite or liberate an electron or electrons, or thermally, by generating phonons and raising the temperature of the detector. With the photoelectric effect individual photon events can be registered or the cumulative effect of many photons can be integrated into a current or a voltage or transformed directly into a permanent record such as a developed photographic grain.

The photosensitive material of the detector is chosen to have a photoelectron excitation energy less than, or equal to, the incident photon energy of interest which means that the excitation energy increases linearly with v. Phonons with sufficient energy can also excite photoelectrons and, if the detector is warm enough, phonons in the high energy 'tail' will generate a background signal known as dark current. Roughly speaking, if $hv < 100kT$ (where T is the temperature of the detector and v the frequency of the photons to be detected) the number of phonons in the detector with energies $> hv$ will generate an unacceptably high background signal. This implies that there is a long-wavelength limit to the effectiveness of photoelectric detection for the low light levels encountered in astronomy.

Figure 2.8 shows the relation $hv = 100kT$ (i.e. $T = 144\lambda^{-1}$ for λ in μm). From Figure 2.8 it is clear that for wavelengths longer than approximately 500 nm the detector should be cooled below $0\,°C$. With increasing wavelength the necessary detector temperature drops until at 36 μm the limit of refrigeration techniques is reached at the temperature of liquid helium (4 K).

The curve in Figure 2.8 should not be interpreted too literally since it is possible to detect the brightest infrared sources incoherently at a wavelength as long as 300 μm using thermal detectors at the temperature of liquid helium. In this case the electrical conductivity of the detector is modified by conversion of the absorbed incident radiation to phonons.

Incoherent detectors are usually many-channelled (e.g. photographic plates, CCDs, etc.) and are used to record images in focus or, if single channelled, they are illuminated by the radiation isolated from a single source. In integrating signal generating detectors the accumulated photogenerated charge is sensed as a voltage at the end of an exposure. In photon counting detectors there is sufficient internal gain that individual photon events are detected and recorded. Photon counting and integrating systems are discussed in detail in Chapters 7 and 8.

Coherent detection

At long wavelengths (i.e. $\lambda > 300\,\mu m$) it is necessary to resort to coherent detection methods which exploit the wave nature of the radiation. The heterodyne technique is the most usual. Incoming

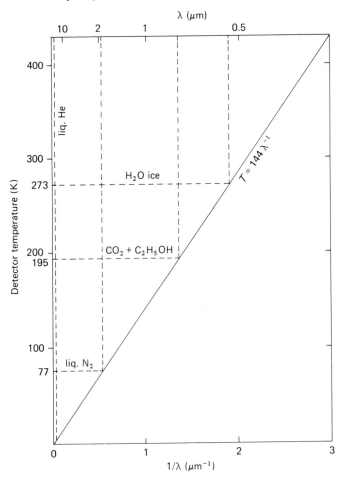

Fig. 2.8 To avoid excessive background noise generated by phonons, a detector should be cooled such that $T < 100k(h\nu)^{-1}$, where $h\nu$ is the photoelectron excitation energy. The plot shows how this limit varies with frequency, and indicates some standard refrigerants.

radiation is mixed with a local oscillator signal and power is measured at the intermediate or beat frequency between the frequencies of the incoming radiation and the local oscillator.

There is no clear demarcation between the infrared spectral regions in which coherent and incoherent detection techniques are used. The choice often depends on the quality of available detectors and on the nature of the observations being made.

Coherent detectors are single channelled and must be fed with radiation of the same phase. At radio frequencies this is achieved by use of an appropriate feed horn (described in Chapter 3).

Detection limits

In any measurement of source flux, radiation is also included from a portion of the sky, parts of the telescope, and, in the case of radio observations, there can be radiation from the ground and aliasing of bright radio sources from side lobes of the telescope response. The background can be natural, man-made, or both, and the shot and wave noise associated with it contributes significant errors. In addition, spurious (usually electronic) signals generated in the detection process appear as part of the signal.

Optical

When a stellar image is detected incoherently the output signal can be expressed as

$$\text{signal} = (nq + Nq + Dm) - (Nq + Dm)' \tag{2.14}$$

where m is the number of detector sensing elements (often called pixels) over which the source is detected (optimum sampling is discussed in Chapter 4), n is the total number of photons from the source delivered to the m detector sensing elements during the observation, N is the number of sky photons delivered to the detector within the m sensing elements during the exposure, q is the responsive quantum efficiency (see Chapter 7) which is the fraction of incident photons detected, and the system readout noise, $D^{0.5}$, (the r.m.s. noise) per sensing element is expressed as D equivalent detected photons (see Chapter 7). Both n and N depend linearly on the exposure time, t, such that $n = (dn/dt)t$, and $N = (dN/dt)t$. D is assumed to be independent of exposure time. $(Nq + Dm)'$ is the average sky and system noise background measured from k resolution elements of m pixels in a star-free area of the image.

The square of the shot noise associated with the signal is

$$\text{noise}^2 = nq + \alpha(Nq + Dm) \tag{2.15}$$

where $\alpha = 1 + 1/k$. If the sky background level can be determined from a large enough area then α can be taken as 1 and the best signal-to-noise ratio which can be achieved is

$$\text{signal/noise} = (nq)/(nq + Nq + Dm)^{0.5} \tag{2.16}$$

Photometric accuracy is normally limited at the one per cent level by calibration uncertainties and variations in atmospheric extinction and sky brightness. These factors tend to be a linear function of $(n + N)$ rather than of $(n + N)^{0.5}$. The number of errors introduced by the passage of cosmic rays through some of the sensing elements increases with exposure time and is discussed in Chapter 8.

The above treatment is appropriate for integrating systems but it is also valid for photon counting except that the system noise can be taken as zero and the responsive quantum efficiency is reduced in the discrimination process.

As an illustrative example, we can compare the theoretical limiting photometric precision possible with three different telescopes, the 3.6 m CFH telescope and the 10 m proposed for Mauna Kea [26], and the 2.4 m Hubble Space Telescope to be launched by the Shuttle [18, 420]. At Mauna Kea the average sky brightness is $m(v) = 22.5 \text{ arcsec}^{-2}$ which corresponds to 30 photons $\text{s}^{-1} \text{ arcsec}^{-2}$ with a 3.6 m telescope and a 50 nm filter bandpass. For the sake of this discussion 50 per cent of the stellar photons collected by the ground based telescope primaries are assumed to lie within 1 arcsec^2 (the effects of atmospheric seeing are discussed in Chapter 4). It follows that stellar magnitude can be expressed as

$$m(\text{star}) = 22.5 - 2.5 \log_{10}(2/r) \tag{2.17}$$

where $r = N/n$, the ratio of the sky and stellar brightness within a circle of 1 arcsec^2.

The sky brightness in orbit for the HST is expected to be 23 magnitudes per square arc second in V which is only half a magnitude fainter than the best ground based values. But, in the absence of atmospheric distortion, the images will be essentially diffraction limited. The individual sensing elements of the Wide Field Camera (which actually undersample the telescope point spread function) have projected dimensions of $0.1 \times 0.1 \text{ arcsec}^2$ on the sky.

From equation (2.16) the general expression for the number of detected stellar photons required to achieve a given signal-to-noise is

$$2nq = s^2(1+r)(1+(1+4Dm/(s^2(1+r)^2))^{0.5}) \qquad (2.18)$$

where the variables are the same as those used in equation (2.16) and s is the signal-to-noise ratio.

The exposure time, t, can be expressed as

$$t = rn(dN/dt)^{-1} \qquad (2.19)$$

Figure 2.9 shows the variation of the log of the exposure time in seconds required to achieve a signal-to-noise ratio of 5 for the range of visual stellar magnitudes 22 to 27. The detector is assumed to be identical in each case using the values of $m = 5$, $q = 0.8$, $D = 100$ (i.e. $D^{0.5} = 10$), and $(dN/dt) = 3 \text{ m}^{-2} \text{ s}^{-1}$. The telescope having the shortest exposure time at a given magnitude is the most effective.

For the ground based telescopes $r \gg 1$ and $N \gg (Dm)^{0.5}$ at the fainter

Fig. 2.9 The variation of limiting V stellar magnitude (signal/noise = 5) with exposure time for a 3.6 m and a 10 m ground based telescope (solid lines) and a 2.4 m satellite telescope (dashed line). The detector ($q.e. = 0.8$, $D^{0.5} = 10$) and filter bandpass (50 nm) are identical in each case. The sky brightness is 22.5 and 23 m arcsec^{-2} from the ground and from orbit, respectively. 50 per cent of the ground based starlight is contained in 1 arcsec^2 and 0.05 arcsec^2 in orbit. See text for further details.

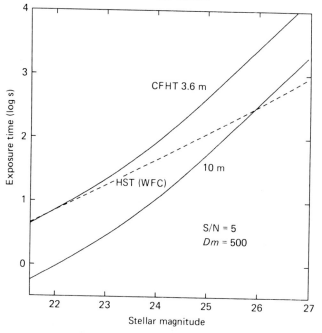

magnitudes. For example, at $m(\text{star}) = 27$, $r > 100$. In this case exposure times are given approximately by

$$t = s^2 r^2 (q\,dN/dt)^{-1} \quad \text{s} \qquad (\text{for } r \gg 1, \; N \gg (Dm)^{0.5})$$

$$= s^2 r (q\,dn/dt)^{-1} \quad \text{s} \qquad\qquad\qquad (2.20a)$$

i.e. t increases as the square of the required signal-to-noise ratio and linearly with r. The difference in exposure times between the 10 m and 3.6 m telescopes is simply the ratio of their collecting areas, 7.7.

The dominant source of noise for the HST WFC is electronic readout noise of the detector since the area of the sky included with the image is 0.05 arcsec^2 compared with the 1 arcsec^2 for the ground based telescopes. D is independent of exposure time and t is given approximately by

$$t = sr(Dm)^{0.5}(q\,dN/dt)^{-1} \quad \text{s} \qquad (\text{for } (Dm)^{0.5} \gg N)$$

$$= s(Dm)^{0.5}(q\,dn/dt)^{-1} \quad \text{s} \qquad\qquad\qquad (2.20b)$$

i.e. t increases linearly with s.

From Figure 2.9 it is clear that a 2.4 m telescope in orbit is equivalent to a 3.6 m telescope on the ground for wide band photometry of stars between about 21 mag and 23 mag. A 10 m telescope would be superior to the HST for stars brighter than about 26 mag. This is one reason why the detection, optically, of stars fainter than 26 mag is an important goal of the HST.

Implicit in the above discussion is the assumption that the number of detected photons, $(n + N)q/m$, does not saturate individual pixels within the image. This is a reasonable assumption for modern signal generating detectors where saturation corresponds to $> 10^5$ detected photons per pixel. But it is not the case for photographic emulsions where dynamic range is very limited.

There are of the order of 10^6 developable grains cm^{-2} for an astronomical emulsion. If $(n + N)q$ exceeds the number of developable grains within the star image, the image will saturate. In consequence, limiting magnitude is a function of the telescope focal length (magnification), the granularity of the emulsion, and the sky brightness. If the stellar image is smaller than a single grain of the emulsion the limiting magnitude increases as the log of the focal length squared (small telescope case), but increases as the log of the focal length when the image is resolved (large telescope case) [52].

Infrared and radio

For both radio and infrared observations, detector noise normally exceeds the noise associated with the signal, and often by a large factor. Detection limits for infrared observations are discussed in detail in Chapters 3 and 7 and for radio observations they are discussed in Chapters 6 and 7.

Fluxes and limit estimates in janskys

Fluxes and detection limits are normally expressed in millijanskys at radio frequencies [114, 235]. It is becoming fashionable to express optical and infrared limits in the same units.

$$1 \text{ millijansky} = 10^{-29} \text{ J s}^{-1} \text{ m}^{-2} \text{ Hz}^{-1}$$

A typical radio bandpass for continuum observations is 50 MHz and, in half an hour with the Very Large Array (VLA), at 20 cm the measured angular r.m.s. fluctuation is a few microjanskys in the background for a blank field which is free of known sources. The VLA is described in Chapter 6. It is insensitive to the d.c. component of the sky background. The associated energy per photon at 20 cm wavelength is approximately 10^{-24} J. The standard deviation of a few microjanskys corresponds to 13 photons r.m.s. detected per second from the approximately 10^4 m^2 of telescope collecting area and the full 50 MHz bandpass, assuming that the measured r.m.s. fluctuation is proportional to the inverse square of the exposure time.

Translating in the opposite direction for optical observations, one has a typical figure of 25 (standard deviation = 5) for the number of detected photons per second from a square arc second of dark sky (equivalent to a sky background of 22.5 mag) with a 3.6 m telescope using a silicon detector and a 50 nm bandpass. The bandpass corresponds to 6×10^{13} Hz at a central wavelength of 500 nm. The telescope primary area is almost 10 m^2 and the energy per photon is 10^{-19} J. The sky background then corresponds to about half a microjansky per square arc second. A limiting stellar magnitude of the Hubble Space Telescope (HST) of, say, 27.5 mag, will correspond to about 5 nanojanskys for the same broad band filter (the HST is described in Chapter 3).

Air mass [179, 361]

The path length traversed by starlight in the atmosphere before it reaches the telescope is known as the air mass and, for stars observed close to the zenith ($z < 60°$), it is given closely by $\tau \sec z$, where τ is the optical depth of the atmosphere vertically above the telescope, and z is the angle of the telescope optical axis to the vertical. $\sec z$ can be expressed in terms of the telescope hour angle and declination (defined later in this chapter)

$$\sec z = (\sin \phi \cdot \sin \delta + \cos \phi \cdot \cos \delta \cdot \cos h)^{-1} \qquad (2.21)$$

where ϕ is the latitude of the observatory, and δ and h are the declination and hour angle of the telescope.

The 'true' air mass is given more accurately by

$$X = \sec z - 0.001\ 816\ 7(\sec z - 1) - 0.002\ 875(\sec z - 1)^2$$
$$- 0.000\ 808\ 3(\sec z - 1)^3 \qquad (2.22)$$

For $z > 80°$, i.e. $X > 5$, a polynomial of sixth order in $(\sec z - 1)$ is more accurate [361].

Optical extinction by the atmosphere

There are three principal atmospheric components which contribute to extinction of transmitted starlight in the optical and infrared spectral regions [186]. The atmospheric scale height and the wavelength dependence are different for each component and, for absolute measurements of the type discussed in the next section, one must solve for each separately [168, 379, 388].

(i) Rayleigh scattering by molecules depends on the temperature profile and pressure at the site. At a standard pressure of 101 325 Pa, and an air temperature of 15 °C, the vertical or zenith extinction caused by Rayleigh scattering, $A_1(\lambda, h)$, in magnitudes/air-mass is well represented by the expression [186, 309]

$$A_1(\lambda, h) = 9.4977 \times 10^{-3} (1/\lambda)^4 (N'/N)^2 \exp(-h/7.996) \qquad (2.23)$$

where N' and N are the refractive moduli of air (see equation (4.2)) at the wavelengths of λ and 1 μm, respectively, λ is in micrometres, and h is the altitude of the observatory in kilometres. Refractive index and, consequently, $A_1(\lambda, h)$, vary predictably with both temperature and atmospheric pressure (see equation (4.5)).

(ii) Aerosol particles, which consist of fine dust, water droplets, and pollution, normally have a smaller scale height than the molecules which cause Rayleigh scattering. The extinction is more 'grey', having a weak dependence on λ. The extinction by aerosol particles, $A_2(\lambda, h)$, in magnitudes per air mass is given by

$$A_2(\lambda, h) = A_0 \lambda^{-\alpha} \exp(-h/H) \qquad (2.24)$$

where H is the scale height, and the values of A_0 and α are peculiar to conditions at the site. Typical values are $\alpha = 0.8$ and $H = 1.5$ km.

(iii) Molecular absorption produces discrete lines and bands. The principal species are ozone (O_3), oxygen, and water vapour.

Ozone is concentrated at altitudes between 10 and 35 km which means that its contribution to the vertical extinction is independent of h for ground based sites. The ozone vertical extinction, $A_3(\lambda)$, is given by [186]

$$A_3(\lambda) = 1.11\,Tk(\lambda) \qquad (2.25)$$

where $k(\lambda)$ is the absorption coefficient in cm^{-1}, and T is the total ozone above the observatory in atm cm, taken from published tabulations as a function of observatory latitude and season [5]. Typical, average values for T are 0.2 at the equator and 0.3 at latitudes of 60°. There are seasonal variations in T and its value can fluctuate significantly during a night.

From the ground, the atmosphere is effectively opaque for $\lambda < 320$ nm because of the ozone absorption.

Figure 2.10 shows a mean variation of the vertical extinction at Flagstaff, Arizona with an indication of the contributions from each of (i), (ii), and (iii) [388]. Not shown are the narrower absorption bands of H_2O at 594 nm, and the A, B, and α bands of O_2 at 762, 687, and 628 nm,

Fig. 2.10 The mean vertical extinction in magnitudes at Flagstaff, Arizona, for 1976 May–June. The assumed contributions of ozone, aerosols, and Rayleigh scattering to the extinction are shown separately. (Published with permission from [388].)

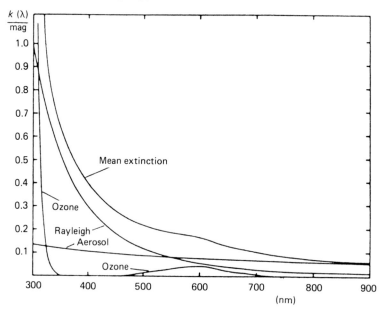

respectively. There are many other lines of these molecules, particularly H$_2$O, which can be found listed in [296].

The amount of water vapour above an observatory is highly variable and it is hard to estimate its contribution in the optical spectrum but it is the dominant source of absorption in the infrared. The atmospheric transmission in the infrared is shown for different amounts of precipitable water above the site in Figure 2.11 [207, 259, 363]. This spectral region can only be observed from the ground at the spectral 'windows' between the absorption bands. Most of these bands are due to water vapour, apart from that near 9.7 μm which is caused by ozone.

The quantity of precipitable water above a site is a function of both the altitude and local climate [272, 280]. On good nights at Mauna Kea

Fig. 2.11 Atmospheric transmission in the infrared between 1 μm and 500 μm for differing amounts of precipitable water above a site. (Published with permission from [259].)

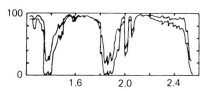

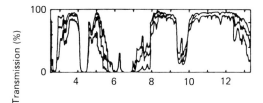

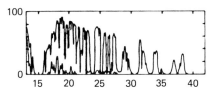

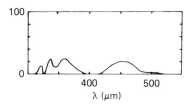

useful observations can actually be made within some of the water vapour absorption bands [280]. For $\lambda > 15\,\mu$m, observations can only be made on the driest nights at high altitude sites. The atmosphere is effectively opaque between 40 and 350 μm.

For differential, broad band photometry, the extinction coefficient for a given filter is determined by observing one or more stars over a range of zenith distances. The gradient of the change of apparent magnitude with X, the air mass, is $k(\lambda)$, the zenith extinction coefficient at wavelength λ. Normally several stars are used covering the range of colours of the programme stars. The effective wavelength of the combination of a broad band filter and the spectral response of the detector is not the same for stars of different colours, nor for stars of the same colour observed at different air masses.

Atmospheric transparency at radio frequencies [41, 235, 411]

Between 0.01 m and 10 m the atmosphere is highly transparent. For $\lambda > 10$ m the ionosphere is highly reflecting and hence opaque. The microwave region is affected by molecular absorption bands principally of O_2 and H_2O in the same way as the infrared. Figure 2.12 shows the

Fig. 2.12 Attenuation of the clear atmosphere in decibels per kilometre with emphasis on the millimetre wave region. Np is the attenuation expressed in nepers where 1 neper is a factor of $1/e$. (Reproduced from [411].)

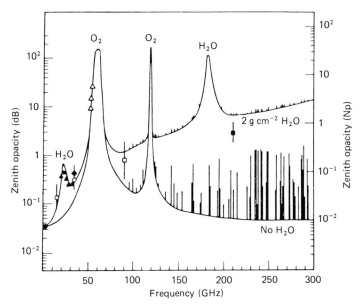

attenuation of radiation in decibels per kilometre by the atmosphere [41, 352]. The plot emphasises the absorption in the mm region. The additional attenuation by rain and cloud is also shown. Rain drops scatter strongly in the mm region with the finer drops in clouds being more effective at shorter wavelengths.

Sky brightness

The radiation which contributes to the brightness of the night sky is both scattered into, and emitted in, the line of sight and it can be separated into

    (i) atmospheric,

    (ii) extra-terrestrial components.

At optical wavelengths, (i) is a mixture of airglow, auroral emission, and artificial and natural light scattered by aerosols. Contributions to (ii) come from sunlight scattered from the Moon and interplanetary dust (zodiacal light and gegenschein), and starlight scattered by interstellar dust grains. The contribution of emission relative to scattering by dust grains increases towards longer wavelengths until emission dominates in the infrared.

The relative importance of each component varies both with position on the sky and with time. The airglow intensity, particularly from the OH bands, varies erratically on time scales of minutes and by up to a factor of two during a night and there are also post-twilight cycles of variation.

A typical spectrum of the night sky between $\lambda = 330$ nm and $\lambda = 1000$ nm at a dark, ground based site is shown in Figure 2.13 [64, 79, 389]. The blue and ultraviolet regions are dominated by weak unresolved O_2 bands and mercury lines from city lights. The latter are more numerous and intense in the ultraviolet where they are more subject to scattering than at longer wavelengths. High pressure sodium lamps have no strong ultraviolet lines but the strong orange lines are accompanied by continua [139].

The red/infrared region is dominated by emission bands of OH and intense auroral lines of O I such as that at 555.7 nm. Moonlight, when present, adds a weak solar spectrum.

If the airglow originates in a uniform, thin layer at height h above the Earth (usually between 100 and 300 km) then, using the van Rhijn method [80, 392],

$$I_0/I(z) = V(h, z) \tag{2.26}$$

where I_0 and $I(z)$ are the airglow intensities at the zenith and at an angle z to the zenith and V, known as the van Rhijn function, is given by

$$V(h, z) = (1 - (a/(a+h)) \sin^2 z)^{-0.5} \tag{2.27}$$

where a is the radius of the Earth. At large zenith angles scattering and absorption act to reduce airglow towards the horizon such that the maximum airglow intensity occurs near $z = 80°$.

Observations in near-Earth orbit are still affected by glow from the upper atmosphere which is partly passive (e.g. geocoronal Lyman-α) and in part excited by the high velocity of the spacecraft. At a relative velocity of 7.6 km s^{-1} the kinetic energies of the various abundant molecular and atomic species correspond to several electron volts which is sufficient to excite a local glow, particularly from O_2.

The absorption and re-emission of radiation in the water vapour and other bands shown in Figure 2.12 make the infrared sky extremely bright, particularly in the region of the bands. The importance of the infrared sky brightness is discussed in Chapter 7.

Radio sky brightness

The daily mean television transmitter power is some 10^7 watts which is greater than any natural source in the solar system apart from the Sun [376]. Although this power is confined to individual broadcast and radar frequencies and their associated side lobes, transmitter power is

Fig. 2.13 A typical night sky spectrum between 330 nm and 1000 nm from a dark, ground based site. (Published with permission from [79].)

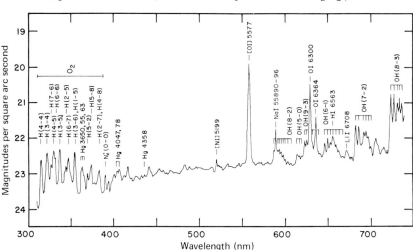

increasing steadily while the radio receivers used by astronomers become increasingly sensitive. A number of frequencies are reserved for radio astronomy and are listed in Table 2.1 [235].

The natural radio background noise is strongly dependent on wavelength and the proximity of the Milky Way. The wavelength dependence is shown in Figure 2.14.

Celestial coordinates

Figure 2.15 indicates how Earth-centred directions are defined in *equatorial* coordinates as well as their relationship to the ecliptic [290]. Declination (δ) is the angle that that direction makes with the Earth's equatorial plane and right ascension (α) is the angle it makes with the plane containing the Earth's axis of rotation and the direction in the sky known as the first point of Aries. The latter is the direction of the Sun at the March equinox.

Hour angle (H) is defined as the angle measured with respect to the meridian plane through the point of observation. The right ascension of a star is thus the difference between the hour angle for the star and the hour angle of the first point of Aries. The hour angle of the first point of Aries (anywhere on Earth) is the Sidereal Time, hence

$$\text{Sidereal Time} = H + \alpha \qquad (2.28)$$

Sidereal Time, α, and H are normally reckoned in hours, minutes, and seconds where 1 sidereal second = 0.997 269 seconds of Universal Time (UT), i.e. sidereal time advances each day by some 3 min 56.6 s of Universal Time. Declination is measured in degrees.

The effect of atmospheric refraction on apparent stellar positions is discussed in connection with telescope orientation in Chapter 3 and for temperature gradients near the telescope in Chapter 4.

Table 2.1 *Frequencies allocated to radio astronomy*

37.25–38.25 MHz	1664.4–1668.4 MHz	10.68–10.7 GHz
73.0–74.6 MHz	1660–1690 MHz	15.35–15.4 GHz
79.75–80.25 MHz	2690–2700 MHz	19.3–19.4 GHz
150.05–153 MHz	3165–3195 MHz	31.3–31.5 GHz
322–329 MHz	4800–4810 MHz	33.0–33.4 GHz
404–410 MHz	4990–5000 MHz	33.4–34.0 GHz
606–614 MHz	5800–5815 MHz	36.5–37.5 GHz
1400–1427 MHz	8680–8700 MHz	

Precession

The rotation axis of the Earth is inclined at 23.5° to its orbital axis and it precesses in a cone of the same semi-angle, 23.5°, with a period of 26 000 years about the orbital axis. This causes the obliquity of the ecliptic to change and there are significant changes in stellar coordinates from year to year. Approximate and precise formulae for the correction of the effect of precession and for the effect of nutation and proper motion can be found in the *Astronomical Almanac* [399]. For back of the envelope

Fig. 2.14 The average variation of natural background radio noise with wavelength. (Published with permission from [235].)

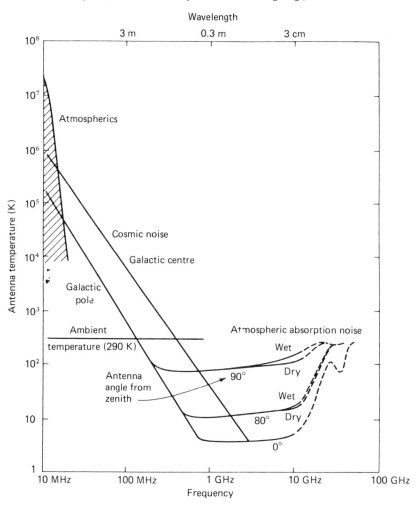

calculations

$$\text{annual precession in } \alpha = 3.0730 + 1.3362 \sin \alpha \cdot \tan \delta \quad \text{s} \quad (2.29a)$$

$$\text{annual precession in } \delta = 20.043 \cos \alpha \quad \text{arcsec} \quad (2.29b)$$

Julian day

In astronomy, times or epochs are normally given in Julian days where the Julian day is estimated from noon (12 hours UT) on 1 January, 4713 BC (JD = 0.0). 1 January, 1980 was JD = 2 444 238, and 1 March, 1990 will be JD = 2 447 952.

Fig. 2.15 The definitions of equatorial and other coordinates discussed in the text (adapted from [290]).

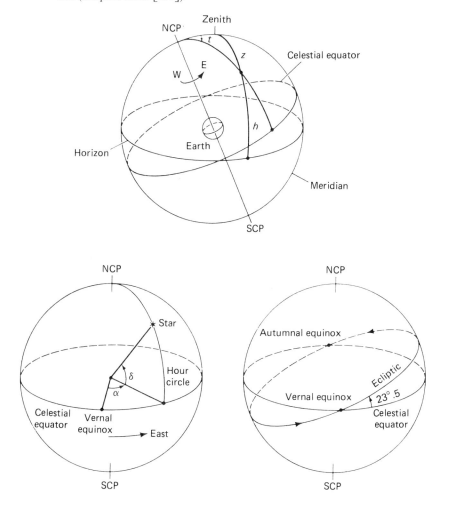

Proper motion

Because of their individual motions in space the apparent positions of stars change with respect to each other. In the case of some nearby stars this can be measured and, as well as providing information about the star's space motion, it leads to errors in telescope setting if not allowed for. The size of the proper motion is designated by μ in arc seconds per year, and its direction, θ, is measured from the direction to the north celestial pole positively through east.

Aberration of starlight

A source appears to be in slightly different directions depending on the velocity of the observer perpendicular to his line of sight. The effect, known as the aberration of starlight, arises from the addition of the velocity vector of the observer to that of the radiation from the source. According to special relativity, the resultant velocity must be c. If α and α' are, respectively, the true (no transverse velocity) and the apparent directions to the source measured from the velocity vector of the observer, as shown in Figure 2.16, then they are related through the formula [281]

$$\tan \alpha' = (\sin \alpha)/(\beta(\cos \alpha + v/c)) \tag{2.30a}$$

or,

$$\tan \alpha = (\sin \alpha')/(\beta(\cos \alpha' - v/c)) \tag{2.30b}$$

where $\beta = (1 - (v/c)^2)^{0.5}$, and v is the velocity of the observer.

The orbital velocity of the Earth is about $30\,\mathrm{km\,s^{-1}}$, hence, a star towards the pole of the ecliptic shows an aberration of some 40 arc seconds over six months. This was first reported by James Bradley in 1729 based on his observations of the star Gamma Draconis and it provided the first conclusive proof of the Earth's orbital motion.

Although equations (2.30a and b) are independent of the velocity of the source, there is a related effect for relative radial motion of the source and

Fig. 2.16 The relationship between the apparent direction, and the source rest frame direction, and the velocity vector of the observer as used in the discussion of the aberration of starlight.

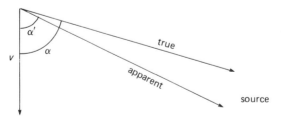

observer. Radiation emitted into the elementary solid angle $d\alpha'$ in the rest frame of the source will lie within the larger solid angle $d\alpha$ of the observer's rest frame, such that [281]

$$d\alpha/d\alpha' = (1 + v/c)/(1 - v/c) \tag{2.31}$$

where v is the velocity (of recession) of the source. When this effect is coupled with the observed increase of recessional velocity with distance for extragalactic sources, the observed, or apparent, diameters of distant clusters of galaxies will not diminish as rapidly as one would expect from the inverse square law. (Space curvature will also affect the apparent diameters of clusters of galaxies at large distances.)

Astrometric precision and astrometric catalogues [121, 226]

Catalogues of stellar and non-stellar objects have been made throughout history. There are over 2000 modern catalogues many of which are redundant, and there is an even larger number of individual lists in scientific publications. The designation and description of sources is quite non-uniform and a given source may appear with a wide range of designations depending on the catalogue.

Accurate coordinates and their variation with time provide three fundamental elements for a source. The precision with which relative positions can be determined depends on how well the centres of individual images can be defined. The most precise stellar positions available in the AGK4 have quoted errors of between ten and twenty milli-arcseconds (the angle subtended by one centimetre at a distance of 1000 to 2000 km). This is sufficient to give accurate trigonometric parallaxes and proper motions for the nearest stars. To achieve this accuracy, all sources of systematic observational error must be carefully calibrated, such as atmospheric refraction, telescope flexure, photographic emulsion shift, personal equation, etc.

It is beyond the scope of this book to explore astrometric techniques in detail. Rather, when discussing individual instruments and detectors some indication will be given of the achievable accuracy.

Fundamental astrometric determinations are of two kinds. Measurement of the position of an object relative to others (usually more distant) within the field of the detector and the estimation of absolute stellar positions. The latter requires measurements over much larger angles. Traditionally this has been done with transit circles [412] and overlapping fields from photographic surveys.

The European Space Agency expects to place the astrometric satellite Hipparcos (HIgh Precision PARallax COllecting Satellite) in a geostationary orbit in 1988. It will have an all-reflecting, eccentric Schmidt telescope (Schmidt telescopes are discussed in Chapter 3). A beam-combining aspheric flat mirror at the centre of curvature will feed the 0.29 m primary mirror with light from two fields separated by a fixed angle of 58° on the sky [29]. A schematic of the optics is shown in Figure 2.17. The telescope focal length will be 1.4 m.

There will be a systematic and continuous coverage of the sky. The satellite will rotate every two hours about an axis normal to the telescope optical axis and at 43° to the solar direction and the rotation axis will revolve about the solar direction once every eight weeks. The telescope will act as a transit instrument measuring the true angle between stars in the two fields.

The two fields are imaged onto the focal plane diaphragm shown in Figure 2.18. The main grid has 2688 slits in a 2.5×2.5 cm area. The slits are 0.47 arcsec wide and on 1.2 arcsec centres. As a star image crosses the slits the modulated light intensity is detected by an imaging photomultiplier. The accurate timing of this modulated signal provides a

Fig. 2.17 Optical schematic of the astrometric satellite Hipparcos expected to be launched in 1988. Two fields separated by 58° on the sky are simultaneously imaged onto the diaphragm (shown in Figure 2.18). The angular separations of stars in the two fields are measured precisely by accurate timing of their transit across the diaphragm with a photomultiplier. Published with the permission of the European Space Agency.

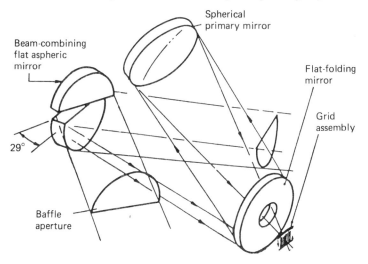

precise position relative to other images. In the expected two-and-a-half-year life of Hipparcos the positions of some 100 000 stars brighter than B magnitude 13 are expected to be measured (about 2 per square degree) with a precision of about 2 milli-arcsec in each of position, proper motion per year, and parallax.

Fig. 2.18 A schematic of the focal plane diaphragm to be used on Hipparcos (see Figure 2.17). Published with the permission of the European Space Agency.

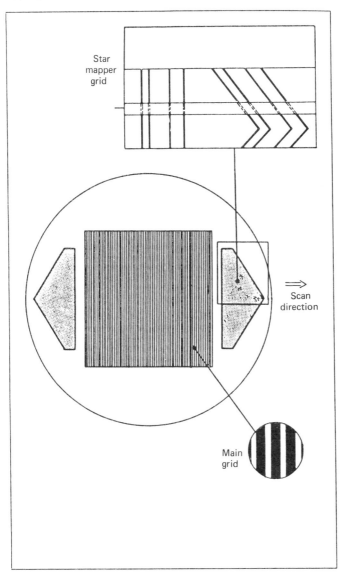

The star mapper grid will provide positions with an accuracy of 0.03 arcsec and B and V magnitudes to 0.05 m precision for some 400 000 stars to a limiting magnitude of 10 to 11 in B.

Doppler shift, radial velocity [160, 161]

In 1842 Christian Doppler published a paper in which he used the analogy of the change in frequency of waves as seen by an observer on a ship in motion compared with one at rest to demonstrate from Huygens' wave theory of light that these same results should apply in astronomy. It was assumed that electromagnetic radiation propagated through an ether and, as with the sea wave analogy, there was a different value for the frequency change depending on whether the source or the observer were moving relative to the ether.

In 1887 Michelson and Morley found that the velocity of light was independent of the orientation of their apparatus to the Earth's orbital motion. In 1905 Einstein, in the special theory of relativity, postulated that the measured velocity of light would be independent of the relative motion of the source and the observer.

The measurement of radial velocities and radial velocity dispersions plays a vital role in studies of the dynamics of individual objects and of systems such as clusters of stars or galaxies. They are fundamental for the determination of masses at all scales, and in the detection of invisible material.

Stellar radial velocities are normally calculated from the relation

$$v = c(\lambda - \lambda_0)/\lambda_0 \quad \text{km s}^{-1} \tag{2.32}$$

where λ is the wavelength measured by the observer for a feature in the stellar spectrum and λ_0 is the wavelength of the same feature in the observer's frame of reference. Radial velocities are positive when the separation of the source and observer is increasing.

In conventional photographic measurements of radial velocity, the stellar spectrum is recorded with the emission lines from a local arc on either side as shown in Figure 2.19. The arc lines, whose wavelengths are known from laboratory measurements, provide a calibration in nanometres of the plate scale, i.e. a dispersion curve. Dispersions are typically in the range of 0.1 to 10 nm mm^{-1}. The positions of selected stellar lines are then calibrated in terms of wavelength and compared with their rest values [312].

The wavelength of a feature seen in a stellar spectrum differs from its laboratory value because of several natural effects such as pressure shifts and blends with other weak lines. Equation (2.32) should be rewritten as

$$v = c(\lambda - \lambda_e)/\lambda_e \quad \text{km s}^{-1} \tag{2.33}$$

where λ_e is the effective wavelength. The wavelengths of the selected lines are not usually the same at each spectral type, nor are they the same at different spectral resolutions. Only lines which are largely unaffected by blends are chosen and the wavelengths adopted are known as *effective wavelengths* [312].

A realistic precision for conventional, radial velocities of bright stars is about 1 km s^{-1}. At 500 nm and a dispersion of 1nm mm^{-1} this corresponds to a displacement of less than $2\,\mu\text{m}$ on the photographic plate.

Autocorrelation techniques

Differential velocities can be measured more accurately by comparing individual stellar spectra from the same star with each other or with the spectrum of a standard star of similar spectral type and known velocity. This is particularly true where displacements are much smaller than the stellar absorption line width and cannot be readily detected by eye in conventional measurement. In autocorrelation techniques an observed spectrum is artificially displaced relative to the standard and the sum of the squares of the residuals between the two spectra is estimated. The displacement with the minimum residual gives the best estimate of the differential velocity.

The radial velocity scanner exploits this technique directly at the telescope [166]. A high-contrast negative widened spectrum of a standard stellar spectrum is used as a focal plane diaphragm in the spectrograph. All of the starlight transmitted by the diaphragm is imaged on a single photomultiplier. The diaphragm can be displaced incrementally and accurately in the direction of dispersion. When the spectrum of a

Fig. 2.19 Part of a conventional photographic stellar spectrum, with the emission lines from a local arc on one side. The latter are used to calibrate distance along the plate in wavelength. Corresponding, displaced neutral iron absorption lines can be seen in the stellar spectrum. Dominion Astrophysical Observatory photograph.

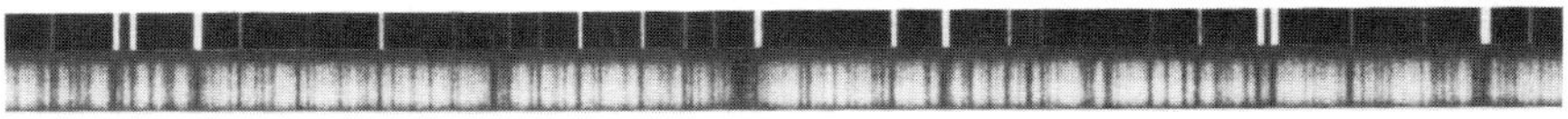

programme star is imaged onto the diaphragm there will be a minimum signal from the photomultiplier when the absorption lines in the programme spectrum are aligned with the transparent slits in the diaphragm.

The radial velocity scanner is an extremely sensitive and accurate device because all of the transmitted starlight falls onto a single detector of high quantum efficiency thereby giving an optimal signal-to-noise. The technique is most effective for cool stars which have a rich spectrum such as the G and K stars.

Where the whole spectrum is recorded, the autocorrelation technique can be carried out analytically on each stellar line separately. If the line profile of interest in the standard is described by the function $p(\lambda)$ and by $p(\lambda + \Delta\lambda_0)$ for the program star then the residuals can be defined by [127, 128]

$$d(\lambda, \Delta\lambda_0) = -\Delta\lambda_0 \, p'(\lambda) + O(\Delta\lambda_0)^2 \qquad (2.34)$$

where $'$ indicates differentiation w.r.t. λ, and $\Delta\lambda_0$ is the true displacement between the program and standard stellar spectra at λ.

When an artificial displacement $\Delta\lambda$ is introduced for the program star, the residuals to first order are given by

$$d(\lambda, \Delta\lambda_0, \Delta\lambda) = \Delta\lambda \cdot p'(\lambda) - \Delta\lambda_0 \cdot p'(\lambda) \qquad (2.34a)$$

If the quadratic function, y, is defined as

$$y(\Delta\lambda) = \sum d^2(\lambda, \Delta\lambda_0, \Delta\lambda) \qquad (2.35)$$

where the summation is over the line profile then, to the lowest order in $\Delta\lambda$ and $\Delta\lambda_0$,

$$y(\Delta\lambda) = (\Delta\lambda^2 - 2\Delta\lambda \cdot \Delta\lambda_0 + \Delta\lambda_0^2) \sum (p'(\lambda))^2 \qquad (2.36)$$

Clearly, $y(\Delta\lambda)$ is a parabola which has a minimum when $\Delta\lambda = \Delta\lambda_0$. The value of $y(\Delta\lambda_0)$ is a measure of the variance in the fit between the standard and candidate profiles and establishes the error in the relative velocity as measured by this technique. The strength of this method is that it uses all of the data points in the line profile and is equally suitable for broad and sharp lines.

Precise differential radial velocities

If wavelength fiducials are imposed in the stellar spectrum by absorption in an intervening medium, systematic errors of conventional techniques introduced by guiding and the lack of common collimation of star and arc light are effectively eliminated [71, 72, 128, 140, 356]. There are several ways in which absorption features can be imposed. The simplest uses telluric lines (i.e. caused by gases in the Earth's atmosphere) [128]. This has the disadvantage that the radial velocity of the intervening atmosphere is unknown, curve-of-growth effects alter the effective wavelengths at different air masses, and it is not possible to obtain spectra of the stars free from telluric lines in order to compensate for the underlying stellar spectrum.

In the other techniques the fiducials are introduced interferometrically [140, 356], or with a gas such as hydrogen fluoride which produces well spaced, sharp absorption lines [71]. A precision of the order of $10 \, \text{m s}^{-1}$ is possible with such techniques [72].

Radio frequency radial velocities

It is possible to measure much smaller fractional changes in wavelength at radio frequencies than at optical wavelengths where (unfortunately) radial velocity is traditionally expressed as

$$v(\text{radio}) = c(v_0 - v)/v_0 = c(\lambda - \lambda_0)/\lambda \quad \text{km s}^{-1} \tag{2.37}$$

this differs from the optical definition and can lead to significant errors in setting frequencies in narrow band radio spectrometers where large optical radial velocities are involved.

Red shift and the K correction [68, 160]

Energetically, the inverse square law is not applicable to sources such as faint galaxies or quasars which have a large radial velocity relative to the observer. From Einstein's special theory of relativity the velocity of light is always observed to be c whatever the velocity of the source relative to the observer. This means that the total photon flux will obey the inverse square law but the observed frequency, and consequently the energy associated with each detected photon, will be modified such that

$$v(z) = v(0)/(1 + z) \quad \text{Hz} \tag{2.38}$$

z is known as the red shift of the source and is defined as

$$z = (v(z) - v(0))/v(z)$$
$$= \Delta\lambda/\lambda_0 = (1 + v \cos \theta/c)/(1 - (v/c)^2)^{0.5} - 1 \tag{2.39}$$

v is the velocity of the source in the rest frame of the observer, θ is the angle of the velocity vector to the line of sight, $\Delta\lambda = \lambda - \lambda_0$.

When v is small and entirely radial to the observer, equation (2.39) reduces to

$$z = v/c \tag{2.40}$$

or

$$v = zc = (\Delta\lambda/\lambda)c \quad \mathrm{m\ s}^{-1} \tag{2.41}$$

which is the form discussed above for stellar radial velocities.

If $\theta = 90°$ then v is entirely transverse to the observer's line of sight and equation (2.39) reduces to

$$1 + z = (1 - (v/c)^2)^{-0.5} \tag{2.42}$$

which is the transverse Doppler shift caused by time dilation.

From (2.39) it follows that

$$d\nu(z) = d\nu(0)/(1 + z) \quad \mathrm{Hz} \tag{2.43}$$

For a 'monochromatic' measurement through a filter of a fixed bandpass this leads to two effects:

(a) a compression of the band width $d\nu(0)$ to $d\nu(z)$ according to equation (2.43),

(b) a change in the effective central frequency of the bandpass $\nu(0)$ to $\nu(z)$ according to equation (2.38).

If the source is non-thermal with a spectral index α (see equation (1.26)) then, from (2.38) and (2.43),

$$f_{\mathrm{observed}} = f_{\mathrm{source}}(1 + z)^{1-\alpha} \tag{2.44}$$

Equation (2.44) is the basis for the K corrections applied to the apparent brightness of quasars of known red shift [68]. If $\alpha = 1$, which is a fairly common value, then $f_{\mathrm{observed}} = f_{\mathrm{source}}$.

Red shift from colour [232]

The large values of z associated with faint galaxies significantly affect their broad band colours. For galaxies fainter than 17 B-magnitude for values of z between 0.3 and 0.6, z can be determined with a standard error of about 0.05 from UBVI photometry. This is quite satisfactory for most cosmological studies and avoids the much less efficient technique of taking a spectrum of each galaxy.

Barycentric and heliocentric corrections

The Earth revolves at 29.8 km s^{-1} in its orbit around the Sun and it rotates with an equatorial velocity of 0.46 km s^{-1}. Consequently there are reflex components of these velocities and other perturbations included in any radial velocity measured from a ground based site. Further, the exact time of observation depends on the point in the Earth's orbit at which the observation was made. In order that times of observation and radial velocities can be directly compared, geocentric observations are normally converted to heliocentric or barycentric values. The former refers to Sun-centred values, while the latter refer to the centre of mass of the solar system. A computer program to convert geocentric to heliocentric values is given by Gordon [161]. For back-of-the-envelope calculations of the heliocentric correction for the two major components one can apply the following formulae

$$\Delta V_0 = 29.80 \cos \beta \cdot \cos(\lambda - \lambda_0) \tag{2.45}$$

and

$$\Delta V_r = 0.463 \cos \phi \cdot \cos \beta \cdot \sin H \tag{2.46}$$

where ΔV_0 and ΔV_r are the corrections for the Earth's revolution and rotation, respectively, for a star with ecliptical coordinates (λ, β). The Sun's ecliptical longitude is λ_0 and can be found in the *Astronomical Almanac* [399]. The latitude of the observatory is ϕ and (λ, β) can be found from

$$\sin \beta = \sin \delta \cdot \cos \varepsilon + \cos \delta \cdot \sin \varepsilon \cdot \cos(\alpha + 90°) \tag{2.47a}$$

$$\sin \delta = \sin \beta \cdot \cos \varepsilon + \cos \beta \cdot \sin \varepsilon \cdot \sin \lambda \tag{2.47b}$$

where ε is the obliquity of the ecliptic as shown in Figure 2.15 which, currently, is about 23.5°. Similar but much smaller corrections can be made for the Earth–Moon barycentre and for the Jupiter–Sun barycentre.

The barycentre is close to the surface of the Sun which revolves around it with a velocity of about $13 \, \text{m s}^{-1}$ in some 12 years largely under the influence of Jupiter's orbital motion. Stumpff has developed programmes which take account of all direct and indirect perturbations by the Moon and planets to the barycentric motion of the Earth. The predictions appear to be accurate to within $0.45 \, \text{m s}^{-1}$ for the velocity and to within 4.6×10^{-5} AU for the coordinates of the barycentre [371, 372].

3

Telescopes

Single telescopes

Astronomers use what are loosely referred to as telescopes to collect radiation from astronomical sources and to estimate their direction and spectral intensities. The wide range of frequencies and photon energies involved in astronomical observations means that the telescopes are very different in form but they can be divided roughly into four categories:

  (i) detectors which sense the direction of arrival and energy of individual (gamma-ray) photons,

 (ii) non-focussing (X-ray) collimators which restrict the field of view of the detector,

(iii) phased arrays, and pencil beam interferometers (metre wavelengths),

(iv) reflecting or refracting telescopes which focus incoming radiation (all wavelengths except gamma rays).

In each category the ability to determine direction and source structure is considerably improved with a 'dilute' aperture of two or more telescopes to increase the baseline. This powerful technique is discussed in Chapter 6. In this section only the properties of single telescopes are discussed, while the detectors are discussed in detail in Chapter 7.

Gamma-ray telescopes – Cos-B [87, 138]

Gamma rays are highly penetrating and consequently cannot be collimated or collected in the same way as lower energy photons. It is also clear from Figure 2.5 that the flux of gamma-ray photons must be very small even from a highly energetic source. In fact, the background of cosmic-ray particles within any energy interval exceeds the integrated flux from the known gamma-ray sources by a factor of the order of 10^5 [257].

Attempts have been made to detect high energy gamma rays from the Čerenkov radiation emitted from high energy electrons generated in the

Earth's atmosphere (see Chapter 1 and [130]). There is a considerable, successful, history of cosmic gamma-ray detection from balloon altitudes but the illustrative example given here is of a satellite-borne instrument.

Figure 3.1 is a sectional view of the Cos-B satellite (Cosmic Ray Satellite-B) which was launched in August 1975 to detect gamma-ray photons with energies >20 MeV [28, 413].

There is a 16-gap wire-matrix spark chamber (SC) and interleaved between the gaps are 12 tungsten plates in which gamma-ray photons tend to produce electron–positron pairs in the Coulomb fields of the tungsten nuclei. The threshold energy for pair production, E, is given by the expression

$$E = 12.5 m_0 c^2 (1 + m_0/M) \cdot 10^{18} \quad \text{eV} \tag{3.1}$$

where M is the nuclear mass associated with the Coulomb charge, and m_0 is the electron rest mass. The high Z of the tungsten minimises energy loss by the electron pairs through nuclear recoil for a given gamma-ray conversion probability.

Normally the electron and positron do not emerge with the same energy and their angle of separation is a function of the energy of the

Fig. 3.1 A sectional view of the Cos-B satellite which detects gamma-ray photons with energies >20 MeV. The spark chamber, SC, and the other components are described in the text. (Published with permission from [413].)

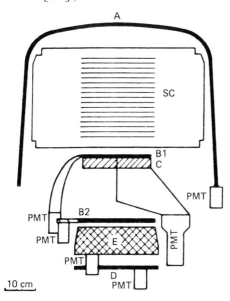

incident photon. The electron–positron pairs ionise gas in the chamber which induces a spark in the high voltage layers of the wire-grid structure. A current flows in the nearest, orthogonal wires thereby setting associated magnetic cores and providing x, y coordinates for the electron or positron position at that particular level in the chamber. In this way the electron pairs can be tracked. Figure 3.2 shows a gamma-ray pair production event as seen in the orthogonal x, y views.

To minimise the frequency with which cosmic-ray particles trigger the system the plastic scintillator, A, acts as a guard. It scintillates on the passage of a charged particle and it operates in anticoincidence to the coincidence triggers from scintillation counters B_1 and B_2 and the directional Čerenkov counter C. The triggering of B_1, B_2, and C defines a

Fig. 3.2 In Cos-B, electron–positron pairs are produced by gamma rays passing through tungsten plates interleaved in the spark chamber. They are detected by wires in the spark chamber which provide x, y coordinates at each level. This figure shows views of a pair production from each of the x and y directions. The angle of separation is a function of the energy of the incident gamma-ray photon. Published with permission of the European Space Agency.

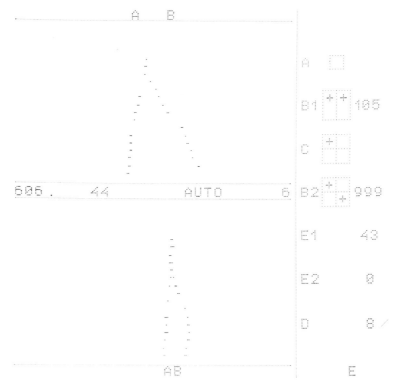

30° field of view for the experiment (full width half maximum). E is a caesium iodide scintillator which measures the original photon energies by absorbing the secondary particles. Scintillator D detects particles from the most energetic events which are not completely absorbed in E.

A proportional counter to measure X-rays in the 2–12 keV range is mounted alongside the gamma-ray telescope to provide synchronism for possible short period pulsations of gamma-ray emission from certain sources known to pulsate in X-rays. The satellite is spin-stabilised at about 10 rpm about its axis of symmetry. The satellite is in an eccentric orbit (apogee $= 9.4 \times 10^4$ km, perigee $= 5.7 \times 10^3$ km, inclination 94.5°) to minimise the time spent in the Earth's radiation belts where the high voltages must be shut down to avoid damage to the photomultipliers.

The satellite is spin-stabilised about the optical axis of the experiment and its pointing direction can remain fixed for as long as necessary to complete an observation. Observations of about one month usually give adequate statistics over a field of some 0.3 steradian [277]. For example, in one month about 3000 photons (> 70 MeV) were collected from an intense region of the Milky Way, while about 300 were seen from a source like the Crab Nebula and its pulsar.

Table 3.1 lists the discrete gamma-ray sources found in the first 18 months of operation together with their flux levels. The diffuse galactic background is discussed in Chapter 1. The instrumental background exceeds the cosmic gamma-ray background by some two orders of magnitude. In the first five years of operation there were over 1.5×10^7 spark chamber triggers which yielded about 10^5 useful gamma-ray events.

The principal sources of uncertainty in defining the direction of a gamma ray are:

   (i) measurement accuracy of a spark chamber track,
  (ii) multiple scattering of the electron and positron,
 (iii) correct identification of the electron and the positron in both projections,
 (iv) correct assignment of a weighted bisector to the electron–positron tracks (the gamma-ray energy is not divided equally between the electron and the positron),
  (v) mechanical tolerances,
 (vi) spacecraft attitude.

For Cos-B, (ii) and (iv) are the most important. The angular resolution for a single photon in terms of the half width at half maximum (HWHM) in the distribution of the photon arrival directions is 3.7° at 100 MeV and about 1° at 1 GeV.

Table 3.2 lists the gamma-ray satellite missions, both launched and planned.

Non-focussing collimators

Collimators of this type restrict the field of view of the detector mechanically and, although now used mostly for X-rays, they can be used for any spectral region (except gamma rays) where large area detectors are available [154]. A slat collimator is shown schematically in Figure 3.3. It is made up of rectangular slots each with an opening of cross-section $a \times b$ and height h. If the radiation of interest cannot penetrate the slat walls then the collimator has a triangular response, such that the half angles of transmission (full width half maximum) in the orthogonal directions are

$$\tan \theta = a/h, \qquad \tan \phi = b/h \tag{3.2}$$

Mechanical collimators are not usually made with half-transmission angles smaller than about 0.5° because sensitivity has to be sacrificed as the angular resolution is improved. In scanning a solid angle α on the sky the fractional dwell time of the telescope on the source is α/Δ where Δ is the collimator field of view.

The modulation collimator, of which a two-wire version is shown in Figure 3.4, overcomes the problem of limited dwell time [154, 306]. Two parallel planes of wires separated by a distance d, with the wires separated by their own diameters, a, provide an angular resolution of a/d in the plane normal to the wires. A source with an angular size similar to, or

Table 3.2 *Gamma-ray astronomy satellites*

Explorer II	USA	1961	gamma rays first detected
OGO-5	USA	1968	
Vela 5a, b; 6a, b	USA	1969	first detection of bursts
Venera 11, 12	USSR/France	1978	detection of bursts
Venera 13, 14	USSR/France	1981	detection of bursts
Prognoz 8, 9	USSR/France	1977–80	detection of bursts
SAS-2	USA	1972–73	first all-sky survey
TD-1	ESRO	1972–74	
Cos-B	ESRO/ESA	1975–82	see text
Planned			
Gamma Ray Obs.			
(GRO)	USA	1988	

Table 3.1 *Gamma-ray sources observed by Cos-B*

Position (1950.0)		Galactic		Intensity $(70 < E < 2000$ MeV) photon cm^{-2} s^{-1}	Counterpart
RA	Decl.	l^{II}	b^{II}		
$20^h\,24^m$	$+39°$	$78° \pm 2°$	$+1° \pm 2°$	$(3 \pm 1) \times 10^{-6}$	
$22^h\,44^m$	$+64°$	$110° \pm 2°$	$+5° \pm 4°$	$(2 \pm 1) \times 10^{-6}$	
$2^h\,44^m$	$+61°$	$135.5° \pm 2°$	$+1.5° \pm 2°$	$(2 \pm 1) \times 10^{-6}$	
$5^h\,6^m$	$+28°$	$176.6° \pm 0.6°$	$-6.9° \pm 0.6°$	$(2.1 \pm 0.6) \times 10^{-6}$	
$5^h\,34^m$	$+22°$	$184.7° \pm 0.3°$	$-5.3° \pm 0.3°$	$(5.0 \pm 1.3) \times 10^{-6}$	PSR0531 + 21
$6^h\,4^m$	$+21°$	$189.5° \pm 1.0°$	$+0.2° \pm 1.0°$	$(3.4 \pm 0.9) \times 10^{-6}$	
$6^h\,32^m$	$+17°$	$195.9° \pm 0.5°$	$+3.6° \pm 0.5°$	$(3.8 \pm 1.0) \times 10^{-6}$	
$8^h\,34^m$	$-45°$	$264.0° \pm 0.5°$	$-2.8° \pm 0.5°$	$(1.1 \pm 0.1) \times 10^{-5}$	PSR0833 − 45

smaller than, a/d is detected with maximum intensity each time the direction of the source is an integral multiple of a/d.

A rotating modulation collimator has a sawtoothed response to a point source while sources of larger angular extent produce no modulation.

Fig. 3.3 Schematic of a slat collimator. The walls of the rectangular slots limit the view of the underlying detector to the half angles $\theta = \arctan(a/h)$ and ϕ. The response pattern is triangular in each direction. (Published with permission from [154].)

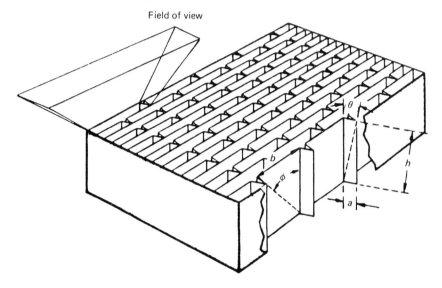

Fig. 3.4 Cross-section of a two-grid modulation collimator. The small circles are wires arranged in two planes which are separated in each plane by their diameters, a. Radiation from a point source on the left is alternately occulted or detected as the system scans across the sky producing a sawtoothed output. On the right, radiation from an extended source always reaches the detector and does not produce a modulated output. (Published with permission from [154].)

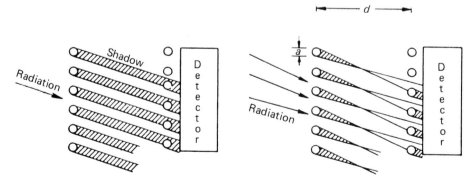

Thus angular resolution and sensitivity are gained at the expense of directional ambiguity and insensitivity to extended sources.

Figure 3.5 shows an outline of the SAS-1 or Uhuru (not an acronym but Swahili for freedom) X-ray satellite observatory which was launched in December 1970 and carried out an all-sky survey in 1971 for X-ray sources in the 1.7 to 18 keV range [155]. Two sets of proportional counters point in opposite directions and each is preceded by a mechanical collimator. The proportional counters are described in Chapter 7.

The collimators are aluminium tubes with 0.25 mm thick walls. One set has a cross-section of 12.7×12.7 mm^2 to give a $5° \times 5°$ field and the other

Fig. 3.5 Major elements of the Uhuru X-ray satellite observatory which carried out an all-sky survey in the 1.7 to 18 keV range in 1971. (Published with permission from [155].)

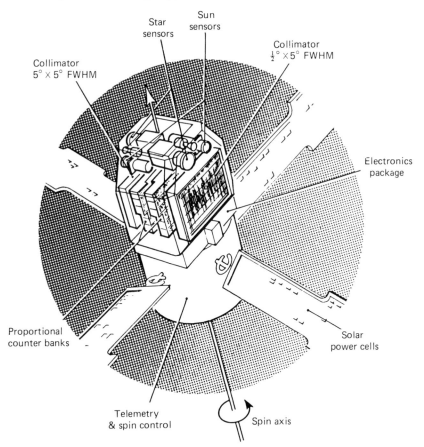

has a 1.27×12.7 mm^2 cross-section for a $\frac{1}{2}° \times 5°$ field (FWHM). Figure 3.6 shows the variation of the mass absorption coefficient of aluminium as a function of photon energy [154]. The rapid drop in absorption above the K-shell absorption energy limits the effectiveness of the collimators to energies of less than 10 and 20 keV for the wide and narrow collimators, respectively. There are different limits because off-axis X-ray photons would have to penetrate some 100 wall thicknesses to reach the detector in the narrow collimator case, whereas the number is more like 10 for the wide collimator.

The spacecraft rotated once every 720 seconds and the spin axis could be changed on command from the ground. The scanning geometry is shown in Figure 3.7. Two identical optical sensors are rigidly mounted parallel to the direction of each collimator. Each focusses images of

Fig. 3.6 Mass absorption coefficient of aluminium showing the K-shell absorption edge at 1.56 keV. (Published with permission from [154].)

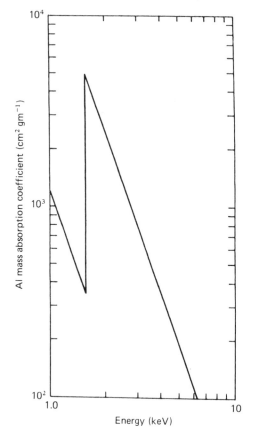

transiting stars on to an N-shaped slot in front of a photomultiplier. The spacings of the triple peaked signals caused by the transit of the star images provide the necessary aspect information.

Phased arrays

At metre wavelengths, telescopes usually consist of dipole arrays which can be steered to respond to radiation from a particular direction by introducing appropriate phase delays between the dipoles before their outputs are combined. When the array is in the form of a cross or a T, appropriate multiplication of the dipole outputs produces a pencil beam [39, 294, 339]. The same principle is also used at shorter wavelengths where, instead of dipoles, the individual elements are radio dishes.

A half-wave or full-wave dipole (such as that shown in Figure 3.11) is an effective absorber of radiation polarised parallel to the dipole over an area defined roughly by the length of the dipole and a width of $\lambda/2$ centred on it. The exact length of antenna wire used to form the dipole depends on the

Fig. 3.7 Band of sky swept out by the wide and narrow collimators of Uhuru in one revolution of the space craft. The full width of the band is 12.7°. (Published with permission from [154].)

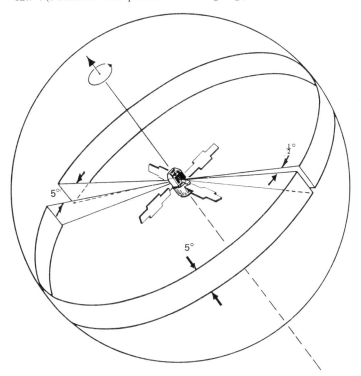

electrical termination used to conduct power to the receiving system [86]. The response of a linear array of parallel dipoles will, to a first approximation, correspond to the classical grating maxima such that

$$Na \sin \theta = d \sin \theta = n \qquad (3.3)$$

where n is zero or an integer, N is the number of dipoles at a distance a apart operating at a wavelength λ, and the full length of the array is d. The spacings of these maxima correspond to those of the double beam interferometer (of separation d) but they are narrower. There are $N-2$ subsidiary maxima or side lobes between the principal maxima which can be suppressed by a suitable grading of the response across the array (see discussion later in this chapter under apodisation, beam tapering, and feed horns).

Fig. 3.8 The connections for a phase-switched two-dipole array. The output signals at A and B show the response of the array to a point source plus a d.c. component of system noise and an extended background source. The signal from B is inverted before being added to A to give the fringe visibility for the point source at C.

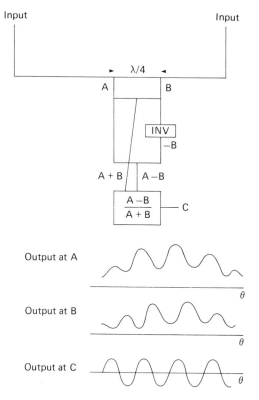

Prior to the introduction of phase-stable correlators, interferometer arrays were normally operated in a phase-switching mode [44, 340]. The principle is illustrated in Figure 3.8. The output signals from two antennae are added at points A and B. The phase of one input has been changed by π through the introduction of an additional $\lambda/2$ length of cable. As a result the patterns at A and B are 180° out of phase. Inverting the output from B and forming the ratio $(A - B)/(A + B)$ at C gives the fringe visibility of the source (see Chapter 2) as it drifts through the reception pattern of the array. The d.c. component of the receiver noise level and incoherent radiation from extended sources such as the sky or galactic background are common to both A and B and, consequently, are rejected at C.

Figure 3.9 shows, schematically, a horizontal dipole array built at Cambridge to operate at 3.7 m (81.5 MHz) [196, 197]. The array is optimised for high sensitivity and wide sky coverage. It has 2048 dipoles arranged in sixteen 470 m long east-west rows of 128 dipoles each. The array is 45 m wide and sensitivity is enhanced towards the ecliptic by mounting each row above a tilted reflecting screen. The array is separated, electrically, into two halves and operates as a phase-switching interferometer, and sources are observed during meridian transit.

Phase scanning is achieved by connecting the 16 rows of dipoles independently to a phasing matrix in each half of the array. In the matrix, the signals are combined using a range of phase differences in steps of $\lambda/16$. This provides 16 outputs from the matrix with one for each of the 16

Fig. 3.9 Schematic of the Cambridge 81.5 MHz array which operates on the principle shown in Figure 3.8. The array is divided into two identical halves and acts as a phase switching interferometer. There are 16 rows of dipoles of which only one row is shown. (Published with permission from [196].)

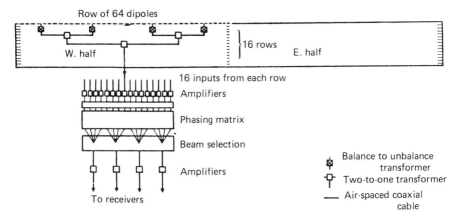

declination angles at which the response is a maximum as shown by the solid lines in Figure 3.10. Intermediate settings in increments of $\lambda/32$ are made in a separate phasing unit to complete the sky coverage as shown by the dashed lines in Figure 3.10.

The decrease in sensitivity at low declinations is caused by a reduction in the projected area of the array and shadowing by adjacent rows.

Pencil beam interferometers

The 22 MHz (13.5 m wavelength) T-shaped array at Penticton, British Columbia, consists of 624 full-wave dipoles, with 368 in the east-west arm, 240 in the north-south arm, and 16 common to both arms [106]. The dipoles are polarised in the east-west direction and a reflecting screen of some 65 000 m² is mounted $\lambda/8$ below the dipoles with the entire structure being supported on a grid of 1698 poles. A cross-section is shown in Figure 3.11.

In the east-west arm the dipoles are arranged in four rows with three groups of 32 dipoles in each and the rows are separated by $\lambda/2$. The north-

Fig. 3.10 The response in declination of the Cambridge 81.5 MHz array for each of 32 possible $\lambda/16$ or $\lambda/32$ phase increments in path length between the rows. (Published with permission from [196].)

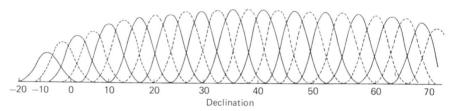

Declination

Fig. 3.11 Cross-section of the 22.25 MHz array showing a dipole and the position of the reflecting screen. (Published with permission from [106].)

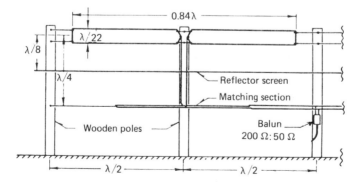

south arm has 64 rows of four dipoles each and, again, the rows are $\lambda/2$ apart. The phasing of the array is achieved by the interconnection scheme shown in Figure 3.12. Solid dots indicate the main phasing for beam 'pointing' and the open circles indicate the rapid $\lambda/2$ switching for phase switching. A hybrid network allows the dipoles in the overlapping region to be shared by both arms.

Figure 3.13 shows the tapering or excitation functions applied to the two arms of the array in order to minimise side lobes. The instrument has a pencil beam of 1.1×1.7 degrees (FWHM).

Reflecting telescopes [22, 25, 55, 286]

Parabolic reflectors are part of the more general family of conic reflectors formed by rotating conic sections about their axes of symmetry. Any conic reflector has a pair of foci or conjugate points on its optical axis or axis of symmetry such that a point source of radiation placed at one is perfectly re-imaged at the other by reflection in the conic surface [402].

Fig. 3.12 The phasing connections for the 22.25 MHz array. Solid points are for beam pointing. The open circles for $\lambda/2$ phase switching. (Published with permission from [106].)

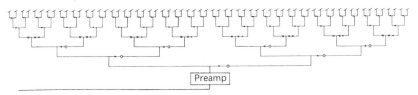

Fig. 3.13 The tapering or excitation functions applied to the two arms of the 22.25 MHz array in order to minimise side lobes. (Published with permission from [106].)

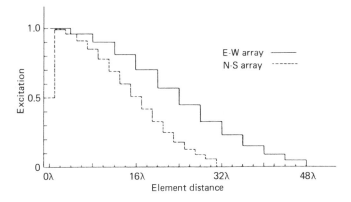

The sphere is unique since its foci coincide at the centre of curvature, a property which is exploited in the wide angle Schmidt telescope (discussed below) where the aperture stop is placed at the centre of curvature of the primary. (The foci corresponding to the conjugate points should not be confused with the optical focus for light coming from infinity which is at $r/2$.)

One focus of an ellipse lies between the vertex and the centre of curvature while the other lies outside the centre of curvature. An elliptic mirror is used as the primary (with a spherical secondary) in a Dall–Kirkham telescope and as the secondary (with parabolic primary) in a Gregorian telescope. In the latter the prime focus coincides with one focus of the ellipse while the other provides the working focus for the astronomer. Both systems are shown in Figure 3.14.

The parabola can be considered as an extreme case of an ellipse in which one focus moves to midway between the vertex and the centre of the vertex sphere, while the other focus is at infinity. (The radius of the vertex sphere is twice the focal length of the paraboloid and has its centre on the optical axis of the paraboloid.) Stars are effectively at infinity and a paraboloid transforms plane-parallel wavefronts incident along the optical axis to spherical wavefronts which converge to a point. A Newtonian telescope which uses a paraboloidal primary is shown in Figure 3.14.

A convex hyperboloid is frequently combined with a paraboloid to make a Cassegrain or a coudé telescope. The former is shown in Figure 3.14. One focus of the hyperboloid lies between the vertex and the point midway between the vertex and the centre of curvature of the vertex sphere while the other focus is virtual and lies behind the vertex. When

Fig. 3.14 Four common telescopes using conic reflectors.

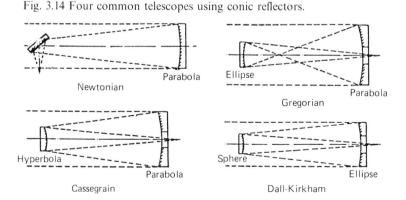

used with a paraboloid, the internal focus of the hyperboloid is put at the focus of the parabolic primary, with the reflected radiation being focussed at the virtual focus of the hyperboloid.

There are some first order relations which are useful for back-of-the-envelope calculations [150]. All of the distances shown in Figure 3.15 are positive (i.e. the secondary is concave). The focal lengths of the primary and secondary are f_1 and f_2, respectively. They are a distance d apart, and the focal plane of the telescope lies a distance e behind the vertex of the primary. The Cassegrain focal length f is given by

$$1/f = 1/f_1 + 1/f_2 - d/f_1 f_2 \tag{3.4}$$

(note that f_2 would be negative for a convex secondary) and

$$(f + f_1)d = f_1(f - e) \tag{3.5}$$

$$-f_2(f - f_1) = f_1(d + e) = f(f_1 - d) \tag{3.6}$$

The Cassegrain focal ratio is simply f/D, where D is the diameter of the primary and the plate scale is $2 \times 10^5/f$ arcsec mm^{-1} for f in mm.

The amplification of the focussing motion of the secondary at the Cassegrain focus is given by

$$\Delta e = (1 + m^2)\Delta d \tag{3.7}$$

where $m = f_1/f_2$ and Δe is the change in position of the Cassegrain focus for a motion Δd of the secondary relative to the primary. The radius of curvature r_{Cass} of the Cassegrain focal plane is given by

$$1/r_{\text{Cass}} = 1/f_1 + 1/f_2 \tag{3.8}$$

The curvature is normally similar to that of the secondary since it usually has the smaller radius of curvature.

Fig. 3.15 Sign convention and dimensions of the two-mirror telescope discussed in the text [150].

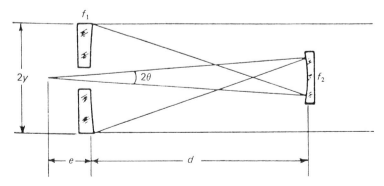

Telescope exit pupil

It is important to know the position of the telescope exit pupil in order that the optics and detectors of auxiliary instruments can be properly matched to the telescope optics [182]. For accurate photometric and astrometric measurements the light from each star being observed, whatever its position in the focal plane, and the light fed in from local sources for calibration should fill the instrumental entrance pupil with the same distribution of intensity. This is particularly important for wide field instruments and for photometric calibration. Where there is no such pupil, as in direct photography at the prime focus, the telescope exit pupil also acts as the instrumental entrance pupil.

The entrance pupil of the telescope is normally the primary mirror aperture or, in the case of a Schmidt-type telescope, it is the aperture of the corrector plate. The exit pupil lies a distance l behind the secondary where [182]

$$l = f_2 d/(f_2 - d) \qquad (3.9)$$

With a concave secondary, l is negative lying in front of the secondary, and behind it if the secondary is a hyperboloid, i.e. f_2 negative. The diameter of the exit pupil p is Dl/d.

Figure 3.16 illustrates a simple matching of the telescope exit pupil to the pupil of the eye. When examining a star image with an eyepiece the image of the primary formed by the eyepiece on the pupil of the eye must not exceed the diameter of the pupil of the eye if light is not to be lost by vignetting of the beam. The diameter of the exit pupil is

$$p = D f_{ep}/f \qquad (3.10)$$

where f_{ep} is the focal length of the eyepiece. For the dark-adapted eye p is about 7 mm, hence

$$m_{min} = f/f_{ep} = D/7 \qquad (3.11)$$

Fig. 3.16 Matching of a telescope exit pupil to the entrance pupil of the eye.

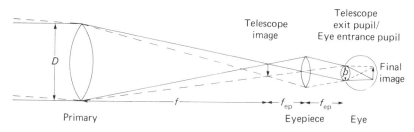

for D in mm. This means that m_{min}, the minimum magnification, increases directly as the telescope diameter.

Under normal daylight conditions the eye has an angular resolution of about one-tenth of a degree [222]. In adapting to the dark individual rods reconnect into groups and the resolution becomes closer to one degree. To just resolve a one arc second star image on the retina of the dark adapted eye therefore requires a magnification m of 3600. Hence, $f/f_{ep} = 3600$, and from equation (3.11) $D_{optimum} = 25.2$ m! Single aperture telescopes of this size might still be realised in this century.

Reflectivity

A single paraboloid or paraboloid–hyperboloid combination can be used for any spectral region for which material of sufficiently high reflectivity is available. The dimensions of the telescope are only restricted by the figuring tolerances on the surface and these are directly proportional to the wavelength (usually $\lambda/10$).

At radio frequencies, continuous surfaces made from available construction metals have a high enough conductivity to give essentially 100 per cent reflectivity. This is also true of thin sprayed-on metallic layers. In general, to reduce weight and cost, reflectors for all but the shortest radio wavelengths are made of expanded metal or mesh. The reflectivity of a mesh is also very high. Theoretical losses of radiation transmitted through wire screens with different values of wire radius, r/λ, and as a function of wire spacing, d/λ, are given in [86]. When $r/\lambda \ll 1$ a wire has high reflectivity only for radiation polarised parallel to its length. For complete reflection the screen must therefore be in the form of a grid.

Figure 3.17 shows a photograph of the 100 metre radio telescope at Effelsberg, West Germany. It can be operated at both prime and Gregorian foci [171], and is the largest fully steerable reflector antenna [136]. The primary focal ratio is 0.3 and the telescope can be used at wavelengths as short as 1.2 cm. The telescope structure is subject to variable, asymmetric gravitational stresses as it points towards different directions in the sky. The rotationally symmetric support adopted for this reflector ensures that, although there are elastic deformations of up to 76 mm, the greatest deviation of the surface from a parabola never exceeds 3 mm. As this is not the same paraboloid in each direction, the prime and Gregorian feeds are deliberately moved to the new focus.

In the optical and infrared spectral regions metallic films are usually vacuum-deposited on to the front, figured, mirror surface. Figure 3.18

shows the reflectivities of freshly deposited films of aluminium, gold, silver, and copper between 200 nm and 10 μm [184]. All are nearly perfect reflectors in the infrared but, except for aluminium, show a sudden loss of reflectivity in the visible or ultraviolet.

Metallic films on ground based telescopes are continuously exposed to the air. This causes silver to tarnish and to lose its high reflectivity unless overcoated for protection. Aluminium has a much higher reflectivity than silver in the important ultraviolet region. It also develops an oxide layer which provides good protection against tarnishing without seriously diminishing its reflectivity.

Multiple dielectric coatings can be used to enhance the reflectivity of a metallic film over a wide spectral range. To obtain the highest reflectivity

Fig. 3.17 A photograph of the fully steerable 100 m diameter radio telescope at Effelsberg of the Max-Planck-Institut für Radioastronomie, Bonn, FRG. The primary is a parabola and the telescope at Effelsberg can be operated at either Gregorian or prime foci at wavelengths as short as 1.2 cm. Max-Planck-Institut für Radioastronomie photograph.

the dielectric layers must be one quarter wavelength thick and be applied in the following sequence: metal, low refractive index film, high refractive index film. Figure 3.19 shows the enhanced blue reflectivity of an aluminium mirror with multiple dielectric layers and a silver based coating which gives enhanced red reflectivity [22]. Such high reflectivities are important when multiple reflections are involved in optical systems such as a coudé mirror train, image slicers, and spectrographs (the coudé focus is described later in this chapter, and spectrographs in Chapter 5).

The aluminium oxide coating reduces the reflectivity of aluminium in the vacuum ultraviolet and the metallic aluminium surface must be protected by a coating of magnesium fluoride or lithium fluoride. Reflectivities for lithium fluoride and magnesium fluoride overcoated aluminium and for uncoated osmium are shown in Figure 3.20 for single reflections. The only advantage of a lithium fluoride overcoat is the access it provides to the important 90 nm to 110 nm spectral region, otherwise magnesium fluoride is to be preferred because lithium fluoride, being hygroscopic and reactive, must be stored in vacuum until launch.

Grazing incidence optics

Although normal incidence reflectivities as high as 8 percent have recently been reported for X-rays from multilayered interference mirrors [390], no single material has a high enough reflectivity at near-normal incidence to justify X-ray optics with the small angles of incidence

Fig. 3.18 Reflectivities of freshly deposited films of aluminium, silver, gold, and copper. (Adapted from [184].)

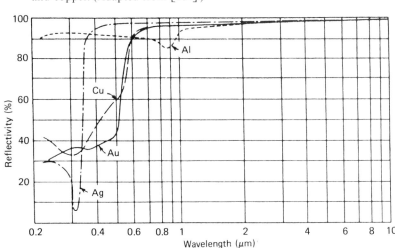

Fig. 3.19 Reflectivities of a multi-layer, all-dielectric coating enhanced for the blue, and a silver based multi-layer dielectric coating enhanced for the red. Dominion Astrophysical Observatory photograph.

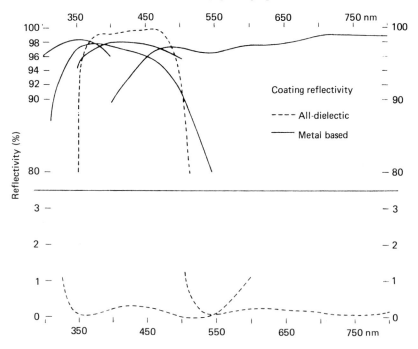

Fig. 3.20 Reflectivities of osmium, and aluminium overcoated with magnesium fluoride or lithium fluoride. From a diagram by G. R. Carruthers.

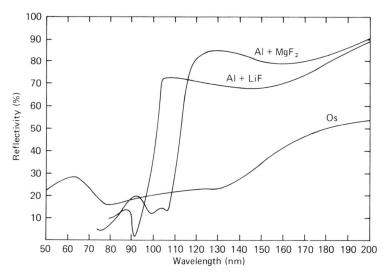

required in Figure 3.14. Fortunately, X-rays suffer some refraction at solid surfaces. The index of refraction is close to, but less than, one, and can be expressed as:

$$\mu = 1 - \Delta \tag{3.12}$$

where Δ is of the order to 10^{-3} to 10^{-6}. This means that X-rays incident on a polished surface at a sufficiently small angle will be totally externally reflected. The critical angle, θ, is given by

$$\cos \theta = \mu \qquad \text{or} \qquad \sin \theta = \theta = (2\Delta)^{0.5} \tag{3.13}$$

Also, for any given substance at wavelengths removed from any absorption edges,

$$\Delta/\lambda^2 = 2\pi r_0 N(e) \quad \text{m}^{-2} \tag{3.14}$$

where r_0 is the classical electron radius and $N(e)$ is the electron density in the material. Hence

$$\theta = 2\lambda(\pi r_0 N(e))^{0.5} \tag{3.15}$$

Figure 3.21 shows the angle of incidence which gives 50 per cent reflection from a silvered surface as a function of the incident X-ray wavelength [422, 393].

Grazing incidence telescope designs for the X-ray and extreme ultraviolet (EUV) spectral regions generally belong to one of four types: the Kirkpatrick–Baez [17], or Wolter types I, II, or III [422, 393]. These are shown in Figure 3.22.

Fig. 3.21 The grazing incidence angle which gives 50 per cent reflectivity from a silver surface as a function of the wavelength of the incident radiation. (Adapted from [422].)

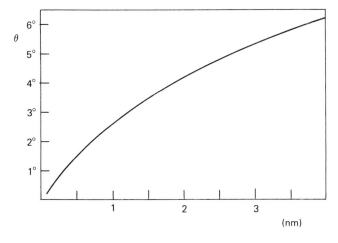

Figure 3.23 is a schematic of a Wolter type II telescope plus detection system which has been flown on a rocket to carry out a sky survey in the wavelength range 10 nm to 100 nm [244]. At 10 nm the critical angle, θ, is close to 10°. The telescope had an effective collecting area of 400 cm², a focal ratio of 4.2, a focal length of 160 cm, with a range of angles of incidence between 3.56° and 6.82°. The off-axis imaging performance is summarised in Figure 3.24 where the dashed line indicates actual performance and the solid line is the theoretical expectation.

Fig. 3.22 Four types of grazing incidence optics for the X-ray and far ultraviolet regions. (Published with permission from [395].)

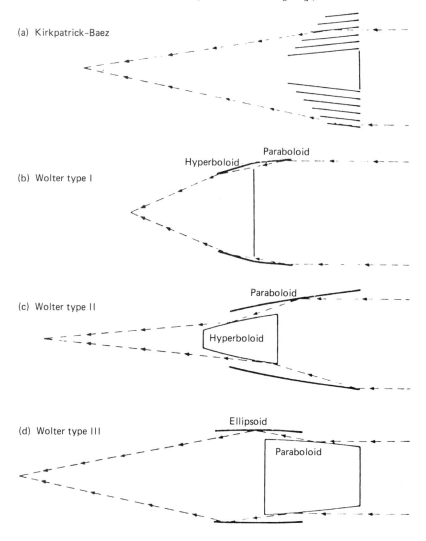

(a) Kirkpatrick–Baez

(b) Wolter type I — Hyperboloid, Paraboloid

(c) Wolter type II — Paraboloid, Hyperboloid

(d) Wolter type III — Ellipsoid, Paraboloid

There were four single paraboloids or concentrators on the SAS-3 satellite [190]. Each fed a single channelled detector and operated at angles of incidence between 2.9° and 4.0° for X-ray energies in the region of 0.25 keV. The proposed Extreme Ultraviolet Explorer Satellite, which may be launched from the Shuttle, will use Wolter type I optics [270, 275].

The Cassegrain telescope on the Einstein Observatory operated between 0.2 and 4 keV with angles of incidence of 70 to 40 arc minutes [153, 292]. The telescope, shown in Figure 3.25, consists of four nested pairs of confocal, coaxial paraboloid/hyperboloids. Quartz was chosen as the mirror material because it best met the goals of one arc second resolution and a minimum of scattered X-rays. The focal length is 344 cm, the diameters of the paraboloids vary between 58 and 33.6 cm, and the

Fig. 3.23 A rocket borne, Wolter type II EUV $f/4.2$ Cassegrain telescope. (Reproduced from [244].)

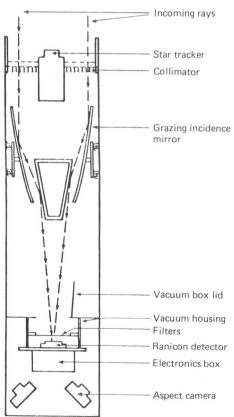

Fig. 3.24 The off-axis image performance of the Wolter II telescope shown in Figure 3.23. (Reproduced from [244].)

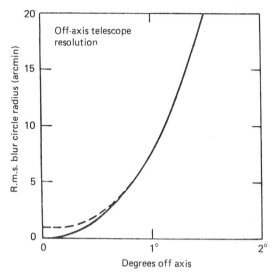

Fig. 3.25 The Wolter type I (Cassegrain) telescope optics of the Einstein X-ray observatory launched in 1978. There are four, nested sets of confocal, coaxial, paraboloid/hyperboloids made of quartz. The diameter of the largest paraboloid is 58 cm and the effective mirror area is 400 cm². Observations were in the energy range 0.2 to 4 keV. (Published with permission from [292].)

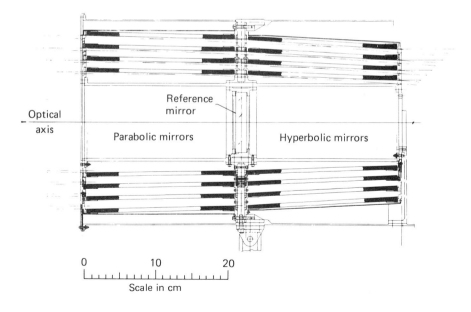

effective mirror area is 400 cm^2. The mirrors are coated with a 25 nm thick layer of chromium and a 60 nm layer of nickel.

The image quality measured by the wide field detector or High Resolution Imager (HRI) [238], which is described in Chapter 8 is shown in Figure 3.26 [394]. The half width at half maximum (HWHM) of a point source image is shown by the lower curve and is valid for all energies. The upper curve gives the radius of the circle containing one-half of the energy reaching the focal plane. The diameter of the Airy disc at 0.25 keV is 1.3×10^{-4} arcsec and therefore does not contribute to the size of the on-axis image which arises from residual surface roughness on the mirrors (r.m.s. values of about 2 nm), slope errors in the figuring (less than 50 nm over lengths of about 2 cm), and alignment errors in the assembly of the telescope. In addition, off axis there are oblique aberrations [282].

Table 3.3 lists the X-ray satellite missions which have been launched and planned at the time of writing.

Oblique aberrations [56, 205]

Although the telescopes in Figure 3.14 give good on-axis images, the image quality deteriorates with increasing angular distance from the optic axis and, if multi-element detectors are used over a wide field, the off-axis images must be corrected.

Fig. 3.26 The off-axis image quality of the Einstein Observatory telescope. (Reproduced from [238].)

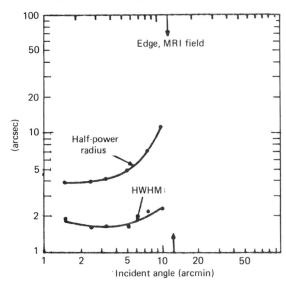

There are five principal aberrations: coma, spherical aberration, astigmatism, chromatic aberration, and field curvature. Coma is the most severe off-axis aberration for simple paraboloidal, or paraboloidal/hyperboloidal, reflectors. Bowen gives the formula [51]

$$L = (5.33)CF^2 \tag{3.16}$$

for an angular distance L from the optical axis at which comatic images reach an angular size C, where F is the telescope focal ratio. This is a purely geometric result and independent of wavelength.

Wide field correction for optical telescopes can be of two kinds. In the first, the field of a classical reflector of the type shown in Figure 3.14 is corrected by transmitting elements placed near the focus. In the second case, the telescope optics are designed as a unit to provide a wide field.

In addition to (negative) coma a paraboloidal mirror suffers from (positive) astigmatism, but is free from spherical aberration. From simple Seidel aberration theory it is not possible to correct spherical aberration and coma simultaneously with a single aspheric plate and, while a thin achromatic doublet lens cannot simultaneously correct for spherical aberration, coma, and astigmatism, Wynne has designed correctors for paraboloids with three well spaced thin lenses which do correct all three

Table 3.3 *X-ray astronomy satellites*

SAS-1 (Uhuru)	USA	1970–71	first all-sky survey
OAO-C (Copernicus)	USA	1972	
SAS-3	USA	1975–79	
Ariel-5	UK	1974–80	
Ariel-6	UK	1979	
ANS	Netherlands	1976	
HEAO-A	USA	1977	
HEAO-B (Einstein)	USA	1978–80	first X-ray images
ASTRO-A (Hakucho)	Japan	1978	
ASTRO-B (Tenma)	Japan	1980	
EXOSAT	ESA	1983	
Planned			
ASTRO-C	Japan	1987	
ROSAT	FRG/USA/UK	1987	
X-ray Timing Explorer (XTE)	USA	1990	
Advanced X-ray Astrophysics Facility (AXAF)	USA	1992	

aberrations as well as providing a flat field and correcting at least in part for chromatic aberration [425, 426]. Figure 3.27 is a schematic of the wide field (1 degree diameter partially vignetted, 46 arcmin unvignetted) blue/visual corrector for the prime focus of the Canada–France–Hawaii 3.6 m telescope, together with a schematic of the corrector mounted in the telescope. The photographs in Figure 3.28(a) and (b) were taken with and without the corrector. The increase of coma towards the edge of the field can be quite clearly seen in Figure 3.28(b).

Schmidt telescopes

While a spherical mirror with a stop at the centre of curvature has no preferred axis and gives images of the same quality over the whole available field it does suffer from spherical aberration. In the Schmidt telescope all orders of the spherical aberration are corrected by introducing an aspherical transmitting element at the centre of curvature [54]. If the refractive index of the transmitting material, usually glass, is n, then the form of the aspheric is given by

$$t = (u^4 - 1.5u^2)D/(512(n-1)F^3) + \text{constant} \quad \text{m} \tag{3.17}$$

where t is the thickness of the aspheric, F is the telescope focal ratio, and $u = h/R$, the ratio of the off-axis distance, h, to the full diameter of the corrector plate, R.

Fig. 3.27 The wide field blue/visual triplet corrector for the prime focus of the Canada–France–Hawaii 3.6 m telescope. It provides a 46 arcmin unvignetted, 1° vignetted field [426]. (Published with permission of Canada–France–Hawaii Telescope Corporation.)

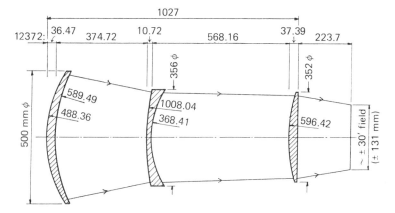

Fig. 3.28 Photographs of two, different 15×15 arcmin fields taken (a) without and (b) with the prime focus corrector shown in Figure 3.28. The field center is at the top left corner in both. The increase in coma with distance from the optical axis can be seen in (a). (Reproduced with permission from plates supplied by R. Salmon of the Canada–France–Hawaii Telescope Corporation.)

(a)

(b)

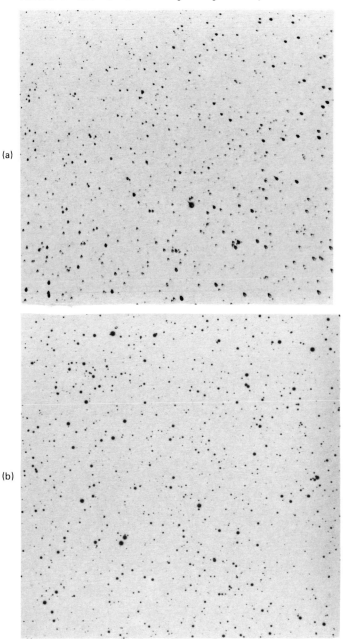

Correction by the aspheric is imperfect for two reasons. First, the refractive index of glass is wavelength dependent which introduces chromatic aberration and, second, the corrector plate described in equation (2.9), unlike the spherical mirror, does have an axis of symmetry and is optimal only for on-axis images. These residual, off-axis aberrations restrict the size of field and focal ratio possible for the telescope [51].

Schmidt telescopes have been used extensively as wide-field survey instruments and the principle has been used widely in the design of fast spectrograph cameras [54].

Figures 3.29 and 3.30 show a schematic and a photograph of the United Kingdom 1.2/1.8 m Schmidt telescope at Siding Spring, Australia [322]. The telescope has a focal length of 3.07 m and its specifications are quite close to those of the Palomar Schmidt. The 1.2 m diameter aspheric corrector is, so far, unique in being an achromatised doublet of Schott-type UBK7 and LLF6 glasses for good ultraviolet transmission.

Ray trace spot diagrams are shown in Figure 3.31. The spots correspond to the intersection with the focal plane of individual rays traced through the optical system. The rays are chosen from across the whole entrance pupil and the spot diagrams indicate the expected angular

Fig. 3.29 A schematic of the United Kingdom 1.2/1.8 m Schmidt telescope at Siding Spring, Australia. Copyright © Royal Observatory, Edinburgh.

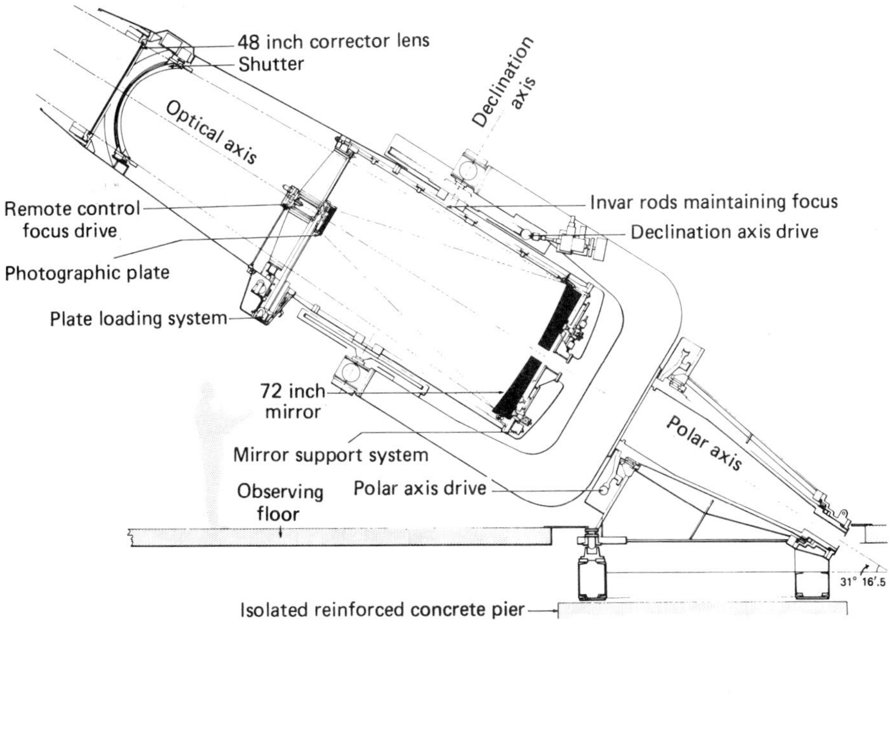

spread which would be introduced by the optics alone. They do not give accurate information about the point spread function. The corrector clearly gives good images from 340 nm to 1 μm over a 6.5° × 6.5° field on 35.6 × 35.6 cm plates [428]. The plates are deformed to the focal plane curvature by a retaining ring and partial vacuum.

Figure 3.32 shows the negative of a photograph taken with the UK

Fig. 3.30 A photograph of the telescope shown in Figure 3.30. Copyright ©
Royal Observatory, Edinburgh.

Schmidt of the Small Magellanic Cloud (SMC) which almost fills the available field together with the globular cluster 47 Tucanae (NGC 104).

Schmidt telescopes have limitations. In order to support the aspheric at the centre of curvature of the primary, the telescope tube must be at least twice as long as a conventional telescope of the same focal length. This requires a dome of twice the diameter which adds considerably to the cost. The focus is inaccessible for use with auxiliary instruments such as spectrographs. Large aspherics can be produced by elastic relaxation techniques but beyond a certain focal length the handling of large

Fig. 3.31 Ray-trace spot diagrams for the achromatic doublet corrector of the UK Schmidt telescope. (Published with permission from [428].)

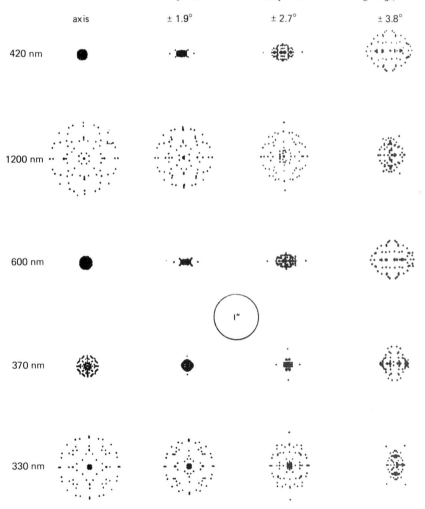

photographic plates to cover, say, a five-degree field becomes very difficult [251]. Further, unless a field flattener is included (which introduces another source of chromatic aberration and ghost images) the plates must be curved to match the spherical focal plane.

The fact that a spherical reflector has no preferred optical axis was exploited at radio frequencies in the design of the 300 m stationary

Fig. 3.32 Photograph of the Small Magellanic Cloud taken with the UK Schmidt telescope. Copyright © Royal Observatory, Edinburgh.

Fig. 3.33 The 300 m fixed spherical reflector of the radio telescope at Arecibo, Puerto Rico. The Arecibo Observatory is part of the National Astronomy and Ionosphere Center which is operated by Cornell University under contract with the National Science Foundation.

reflector built in a volcanic crater at Arecibo, Puerto Rico, and shown in Figure 3.33 [132,258]. The reflecting surface consists of 38 778 thin aluminium panels and the deviation from a sphere is less than 6 mm r.m.s. This stability is achieved by the use of a four-wire suspension at each point of support which is insensitive to variations in temperature [119]. Spherical aberration prevents radiation from coming to a sharp focus and the prime focus feed is, in consequence, a 30 m long 'leaky' cylinder and is shown in Figure 3.34.

Correction of spherical aberration by the use of aspheric secondaries although possible has yet to be used with large spherical dish antennae in radio astronomy.

Since multi-channel detectors are not available at radio frequencies, wide field systems are not exploited except in the stationary, spherical reflectors of the Arecibo type. The great success of dilute aperture techniques for the detailed mapping of radio source structure (see

Fig. 3.34 The 430 MHz prime focus feed of the 300 m radio telescope which has the effect of correcting for the spherical aberration introduced by the spherical reflector. The Arecibo Observatory is part of the National Astronomy and Ionosphere Center which is operated by Cornell University under contract with the National Science Foundation.

Chapter 6) has meant that single dishes tend to be used for the spectroscopy and photometry of individual point sources.

The Ritchey–Chrétien telescope

A large number of optical telescopes have been built in the last decade in the 3 m to 6 m class and a list is given in Table 3.4. Several of these have been designed as Ritchey–Chrétien systems for wide-field photography at the Cassegrain focus [150]. The system is named after the designer Chrétien, and the optician, Ritchey, who first achieved the figuring of the necessary aspheric surfaces.

In a standard Ritchey–Chrétien (RC) system the secondary has a figure which eliminates spherical aberration while the primary figure eliminates coma. This gives the primary a more relaxed figure than a paraboloid by a few wavelengths. The simple RC system suffers from both astigmatism and field curvature but several corrector designs exist to provide good images over flat fields $> 0.5°$ [427].

Figure 3.35 shows an outline of the Hubble Space Telescope (HST) which is to be launched from the Shuttle [18, 420]. It is a 2.4 m $f/24$ RC system which, by virtue of its wide field, will feed six separate, permanently mounted, auxiliary instruments at the Cassegrain focus. The characteristics of the optical system are given in Table 3.5. The expected optical performance is summarised in Figure 3.36 which shows the point spread function (see Chapter 4) and distribution of encircled energy at 623.8 nm and the variation of image quality with off-axis distance.

The central 3 arc minutes of the field are used by the single radial instrument (Wide Field/Planetary Camera WF/PC) while the four axial instruments (Faint Object Camera FOC, Faint Object Spectrograph FOS, High Resolution Spectrograph HRS, High Speed Photometer HSP) view the unvignetted field three or more minutes of arc off axis. Compensation for astigmatism and field curvature is made within the instruments. The outer, astigmatic zone is used by the fine guidance sensors which will also be suitable for astrometric observations of the guide field.

The telescope primary, bulkhead, and baffles will be maintained at 21 °C. Alignment and focus errors will be sensed by a shearing interferometer arrangement and be corrected by actuators on the secondary mirror. (The operation of a shearing interferometer is outlined in Chapter 4.) Errors in the primary figure will also be corrected by actuators on the primary.

Table 3.4 *The largest optical telescopes*

Site	Ownership	Altitude (m)	Hemisphere	Diameter of primary (m)	Year
Mt Pastoukhow, Caucasus, USSR	USSR	2050	N	6.05	1974
Mt Palomar, California, USA	USA	1800	N	5.08	1949
Mt Hopkins, Arizona, USA	USA	2600	N	4.60[a]	1979
La Palma, Canary Islands, Spain	UK	2400	N	4.20	1986
Kitt Peak, Arizona, USA	USA	2100	N	4.01	1973
Cerro Tololo, Chile	USA	2500	S	4.01	1974
Siding Spring, New South Wales, Australia	UK–Australia	1200	S	3.88	1974
Mauna Kea, Hawaii, USA	UK	4200	N	3.80	1979
Mauna Kea, Hawaii, USA	Canada–France–Hawaii Europe	4200	N	3.60	1979
La Silla, Chile	Europe	2450	S	3.57	1976
Calar Alto, Spain	FRG–Spain	2160	N	3.50	1979
Mt Hamilton, California, USA	USA	1300	N	3.05	1959
Mauna Kea, Hawaii, USA	USA	4200	N	3.00	1979

[a] Multiple mirror telescope equivalent light collecting area.

Fig. 3.35 A simplified, cross-sectional view of the Hubble Space Telescope. The telescope is a 2.4 m $f/24$ Ritchey–Chrétien system which will feed six separate instruments in the focal plane. Reproduced from [18].

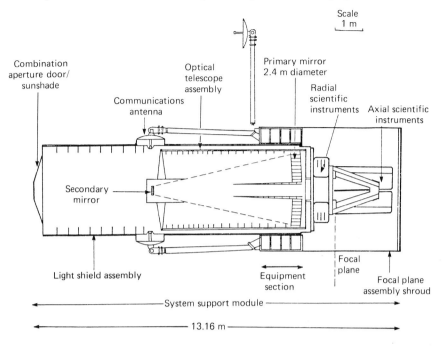

Table 3.5 *Characteristics of the Hubble Space Telescope*

Aperture	2.4 m
Focal ratio	$f/24$
Maximum obscuration ratio (linear)	0.34
Effective focal length (EFL)	57.6 m
Back focal length (BFL)	1.5 m
Primary to secondary spacing	4.90 m
Plate scale	3.58 arcsec mm^{-1}
Field of view diameter	28 arcmin (467 mm)
Data field diameter	18 arcmin (300 mm)
Tracking field size	1.5×10^{-5} sr (180 arcmin)^2
Mirror coating	Al coated with MgF_2
Primary mirror focal ratio	$f/2.3$
Secondary mirror focal ratio	$f/2.23$
Secondary mirror aperture	0.31 m
Secondary mirror magnification	10.4

For direct photography, with any telescope, it is particularly important that the detector have no direct view of the sky. Direct sky light is normally intercepted by baffles, and these are shown in the sectional view of the ST optics in Figure 3.35. The need for, and the extent of, baffling usually restrict the size of the field which can receive light from the whole primary (unvignetted field).

For ground based, multi-purpose telescopes, Ritchey–Chrétien systems have the disadvantage that coma is increased at all foci other than that for which the system was optimised. Although a Cassegrain RC telescope

Fig. 3.36 (a) The expected point spread function of the Hubble Space Telescope, and the distribution of encircled energy in the image at 623.8 nm. (b) The variation of image quality. Reproduced from [18].

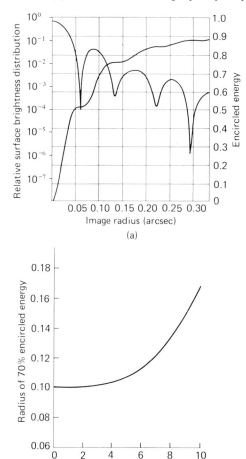

cannot be used for prime focus wide field photography without correctors, the comatic correction in the primary means that the correctors are smaller and simpler (usually two components) than those necessary to correct to a wide field for a paraboloid. Generally the finding field required at the coudé focus for on-axis spectroscopy is about 1 or 2 arc minutes which can be achieved without correction for an RC Cassegrain.

Instrument adaptors/offset guiders

For most large optical telescopes certain basic functions which are common to several instruments are supplied in an adaptor or guider which acts as the mechanical interface to the telescope. It provides rotation and defines the position of the focal plane. The adaptor allows the observer to examine an object or star field before starting an exposure by providing a full-field view within an eyepiece or low light level television system and by acquiring a guide star for use during the exposure. The adaptor often includes calibration sources and the optics necessary to feed light from them into the auxiliary equipment.

Figure 3.37 shows a schematic of a simple adaptor used at the prime focus of the CFH 3.6 m telescope. The field acquisition mirror allows the field to be verified and can be swung out of the way during an exposure. A small probe mirror is used in the focal plane outside the area of the plate to pick up a guide star.

Infrared telescopes

The maximum of the thermal emission at 300 K is near 10 μm. As a result, at wavelengths $> 2.5\ \mu$m, thermal radiation from the sky and from any structures included within the telescope beam contributes a high background signal. Telescopes designed for optical observations are far from optimal for use in the infrared because of the following features [131, 259]:

(i) the Cassegrain focal ratio is too small,
(ii) direct view of the sky is blocked by baffles and all visible surfaces apart from the mirrors are blackened,
(iii) the unvignetted field of view is large and corrected,
(iv) mirror surfaces are coated with aluminium.

Infrared detectors must be operated at low temperatures which means that the telescope entrance pupil should subtend as small a solid angle as possible at the detector in order to maintain proper refrigeration. This condition dictates a focal ratio > 20.

Fig. 3.37 Schematic of the instrument adaptor/guider used at the prime focus of the Canada–France–Hawaii 3.6 m telescope. The prime focus cage outline is shown and the instrument rotator and guiding head are mounted on the wide-field corrector support. The field acquisition mirror can be moved in *x* and *y* and feeds the guiding eye-piece or a television camera (not shown). A small focal-plane probe mirror (not shown) picks up a guide star outside the area of the plate. All dimensions are in mm. (Published with permission from [79].)

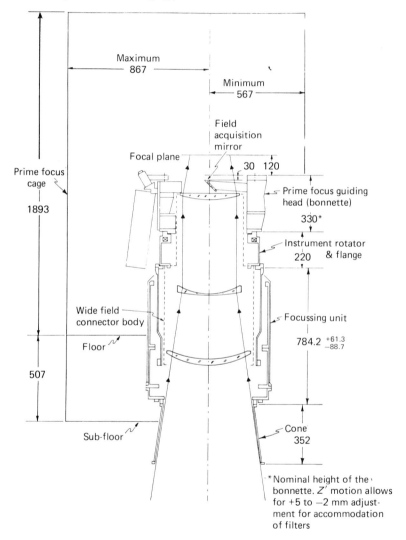

A wide field at the Cassegrain focus requires a large central hole in the primary and a secondary which is oversized for on-axis images. The oversized secondary usually has a broad support frame and tends to deflect unwanted background radiation from around the primary to the focus. Normal aluminium coatings have an emissivity about four times higher than vacuum-deposited silver or gold.

The amount of spurious radiation introduced at $10\,\mu$m by the features (i) to (iv) above is similar to the sky background included with the source. Ground based telescopes specifically designed for observations in the infrared eliminate this additional background by:

    (a) having a secondary which is marginally under the size necessary to give an unvignetted on-axis field,

    (b) use of silver or gold as the reflecting film,

    (c) having a primary with a central hole just large enough for the on-axis Cassegrain beam and substituting skylight with a conical mirror at the centre of the secondary which just covers the image of the hole in the primary,

    (d) introducing refrigerated plugs and baffles to shield all non-reflecting surfaces in the beam,

    (e) using a 'chopping' secondary with a support spider of minimal cross-section.

Item (e) is essential because the infrared sky brightness continually changes and most sources are much fainter than the range of the associated sky fluctuations. By chopping frequently between the source and a nearby portion of the sky and accumulating the difference signal, the sky background and its variations can be eliminated. A chopping secondary can also be used to generate a raster scan of the source [172]. Chopping frequencies range up to 50 Hz and the angular excursions at the focus range up to 2 arc minutes. For the greatest observing efficiency the angular motion between the two 'chopped' positions should be a square wave.

Figure 3.38 shows a photograph of the 4 m UK Infrared Telescope on Mauna Kea, Hawaii, which illustrates many of the points (a) to (e) discussed above [210].

The absence of an atmosphere makes it possible to refrigerate all parts of a satellite telescope without the risk of condensates on the optics. Several cooled telescope designs have been proposed for Shuttle orbit [11, 185, 252, 275, 316, 359].

Figure 3.39(a) shows a schematic of IRAS, a refrigerated telescope which in 1983 surveyed the sky between 8 and 120 μm [103, 247, 375]. The telescope, an $f/9.6$, Ritchey–Chrétien of 57 cm aperature is in a 900 km, circular, twilight orbit. The mirrors are beryllium with the secondary aluminised to enhance visible reflectivity. The telescope is mounted in a superfluid helium dewar. Exterior baffles and a sun shade reduce the heat

Fig. 3.38 Photograph of the 4 m UK Infrared Telescope on Mauna Kea, Hawaii, which illustrates many of the special features discussed in the text. Copyright © Royal Observatory, Edinburgh.

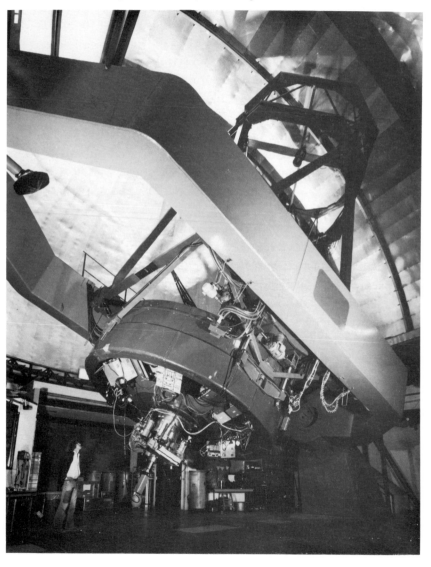

load on the dewar. The telescope optics and dewar wall were maintained below 5 K for the eleven-month holding time of the cryogen. This was sufficient time to complete an all-sky survey.

The survey was made in four wavelength bands centred at 12, 25, 60, 100 μm. The focal plane assembly consisted of a series of rectangular slots followed by an array of 62 detectors operating at 1.8 K. The focal plane diaphragm is shown in Figure 3.39(b) and the characteristics of the focal plane assembly (FPA) are listed in Table 3.6. Either the telescope could operate in a pointed mode or, for the survey, it was pointed radially away from the Earth which provided the best shielding from Sun and Earth heating and gave a natural scan rate. The detectors were so arranged that any source crossing the field was seen by at least two of the detectors in the same wavelength band.

The low resolution spectrograph (LRS) is essentially an objective prism spectrograph which was activated whenever a sufficiently bright source crossed the FPA. The chopped photometric channel (CPC) uses an internal chopper for absolute and differential photometry of selected objects. It simultaneously mapped sources in two bands 41 to 63 μm, and 84 to 114 μm, at a higher resolution than the survey array and its field of view, 1.2 arc minutes, matches the telescope diffraction limit at 80 μm.

Coudé focus

In the spectroscopy of point sources it is often convenient to set up heavy spectroscopic equipment at the coudé focus. This focus has the property that although the field rotates as the telescope tracks a star, on-axis images remain stationary. Light from the telescope is directed along the polar axis either to the north or to the south and, frequently, brought out horizontally to a focus.

With the introduction of the type of high reflection coatings discussed at the beginning of this chapter, it has been possible to minimise reflection losses in the mirror train and, by operating at a high, intermediate focal ratio, to use small secondary mirrors to reduce the central obstruction in the telescope and to limit the cost of the high reflection coatings [324]. The effect of having a lens deliver a smaller focal ratio to the spectrograph puts the telescope exit pupil between the lens and the spectrograph entrance slit. This is extremely convenient for illuminating it with calibration sources. For the most precise spectrophotometry the pupil should be located for each star because it tends to wander with telescope position.

The schematic of a coudé train is shown in Figure 3.40. Lighter auxiliary equipment can be mounted directly onto the telescope at the Cassegrain, prime or other foci.

There are two important shortcomings of the coudé focus: (i) flexure of the telescope structure causes decollimation of the telescope and spectrograph optics, and, (ii) the non-normal incidence of light on the mirrors of the coudé train introduces phase changes which complicate polarisation measurements [14, 48]. These effects depend upon both the

Fig. 3.39 (a) Schematic of IRAS (infrared astronomical satellite), which carried out an all-sky infrared survey in 1983 at 12, 25, 60, and 100 μm [375]. The telescope was refrigerated with liquid helium. (b) The focal plane diaphragm. Courtesy of the Jet Propulsion Laboratory, California Institute of Technology, Pasadena, California.

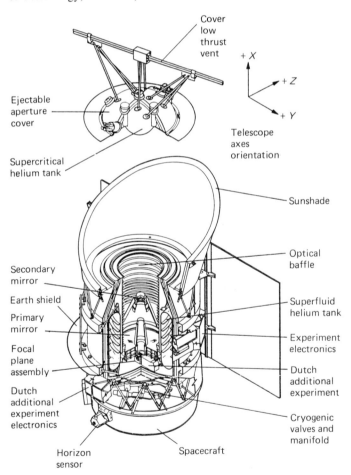

(a)

telescope declination and hour angle and, consequently, are difficult to calibrate.

Telescope flexure

It is not possible to build a completely rigid support for telescope optics. Instead, telescopes are so designed that under gravitational

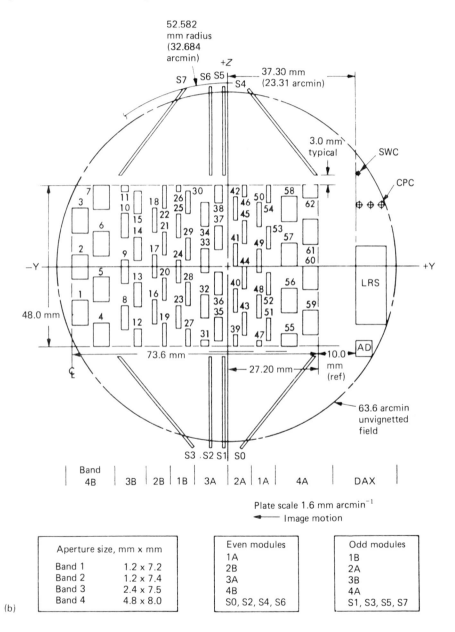

(b)

Table 3.6 *Characteristics of the IRAS focal plane assembly*

	Centre wavelength (μm)	No. of detectors	Detector field of view (arcmin × arcmin)	Wavelength interval (μm)	Detector material	Dwell time (ms)	Resolving power	Average 10σ sensitivity (Jy)
Survey array								
	12	16	0.75 × 4.5	8.5–15	Si:As	190		0.7
	25	15	0.75 × 4.6	19–30	Si:Sb	190		0.65
	60	16	1.5 × 4.7	40–80	Ge:Ga	390		0.85
	100	15	3.0 × 5.0	83–120	Ge:Ga	780		3.0
Chopped photometric channel								
Band 1		1	1.2 (diameter)	84–114	Ge:Ga	1000		7.0
Band 2		1	1.2 (diameter)	41–63	Ge:Ga	1000		7.0
Low resolution spectrometer								
Band 1		3	5.0 (width)	8–13	Si:Ga		14–35	
Band 2		2	7.5 (width)	11–23	Si:As		14–35	

deformation the optical axes of the primary and secondary mirrors remain both parallel and coincident. This can be achieved mechanically by the use of Serrurier trusses as is shown schematically in Figure 3.41 but at the expense of mechanical alignment [22].

The telescope is attached to the polar axis through the centre section which, in turn, supports the secondary and the primary mirrors

Fig. 3.40 Mechanical and optical schematic of the University of British Columbia 0.4 m Cassegrain/coudé telescope, showing part of the coudé mirror train.

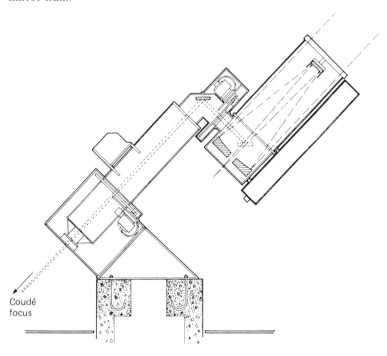

Coudé focus

Fig. 3.41 Schematic of a Serrurier truss support system. Despite gravitational deformation of the telescope structure, the primary and secondary mirrors remain parallel. (Reproduced from [22].)

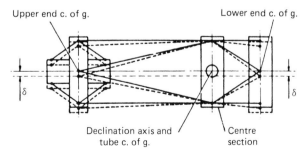

Upper end c. of g.   Lower end c. of g.

Declination axis and   Centre
tube c. of g.   section

independently by Serrurier trusses. The centres of gravity of the mirrors are coplanar with the apices of the trusses with the result that, as the telescope deforms under gravity, the two mirrors remain parallel. For this system to work it is essential that the lateral displacement be identical for both mirrors which in turn requires the stiffness of the trusses to match the relative loads of the mirrors.

The mirrors would also deform under their own weight if they were not properly supported. Modern mirror support systems have become sufficiently refined that primaries with a diameter-to-edge-thickness ratio > 13:1 are possible [210]. The mirror normally 'floats' on a system of supporting levers and, sometimes, radially on a liquid such as mercury [22].

The optical axis of auxiliary equipment at the coudé focus does not share the deflections of the telescope optical axis and, as a result, will not exactly coincide with it. This misalignment can lead to significant wavelength errors in conventional spectroscopy.

Induced polarisation

Oblique reflections from metallic surfaces introduce two effects. If the electric vector of the incident radiation is resolved into a component in the plane of incidence and another normal to it, then the two components are reflected with different efficiencies and phase. The magnitudes of these differences depend on the angle of incidence. In coudé focus polarimetry these effects must be calibrated and corrected [46, 47]. The phase shift, Δ, and the reflection coefficients, r_s and r_p, (where s denotes the component perpendicular to the plane of incidence, and p the parallel component) can be derived (approximately) from the following formulae [48]:

$$\tan \Delta = \sin q \cdot \tan 2p \tag{3.18}$$

$$\cos 2\zeta = \cos q \cdot \sin 2p \tag{3.19}$$

where

$$\tan q = k \tag{3.20}$$

$$\tan p = \eta(1 + k^2)^{0.5}/(\sin \phi \cdot \tan \phi) \tag{3.21}$$

$$\tan \zeta = r_p/r_s \tag{3.22}$$

where η and k are the refractive index and the absorption coefficient of the metal, respectively. The angle of incidence is ϕ and Δ is the phase difference introduced between the p and s components in the sense $\Delta = \delta_p - \delta_s$.

The coefficients η and k are wavelength dependent, with $\eta^2 + \eta^2 k^2 \approx 30$ for aluminium. The phase shifts depend upon the thickness, structure, age, and cleanliness of the metallic coating. Table 3.7 gives some experimentally determined values of Δ (as a fraction of λ) at two wavelengths over a range of incident angles for a freshly aluminised surface [48]. A Babinet–Soleil compensator [45] is normally used to correct optically for the accumulated phase shift in the mirror train [48].

One serious drawback of the multi-layer dielectric coatings mentioned earlier is a catastrophic depolarisation which they introduce at non-normal incidence. This makes them quite unsuitable for polarimetry at the coudé focus [46]. This problem appears not to be shared by overcoated metal mirrors.

Spectrograph slits also introduce significant amounts of polarisation. For example, a 20 μm wide steel slit will induce 30 per cent polarisation in unpolarised light at 1 μm [223].

Optical fibres [20, 200, 263, 315]

The high transmission efficiency of some available quartz fibres makes them suitable as optical waveguides either in the detailed manipulation of light at the focal plane or to carry light many metres from a telescope focus to stationary instruments [7, 189]. Irregularities in the fibres tend to reduce the effective focal ratio of the emerging beam relative to its input value, particularly for large (> 10) input focal ratios. From the discussion in Chapter 5 on spectrograph efficiency, it is clear that any artificial reduction of the collimator focal ratio will reduce spectrograph efficiency.

Table 3.7 *Phase shifts measured after reflection from fresh aluminium*

	Phase shift (in λ)	
Angle of incidence	630 nm	450 nm
40°	0.47	0.46
45°	0.465	0.45
50°	0.46	0.43
60°	0.43	0.39
70°	0.385	0.32

Figure 3.42 shows experimental results for a particularly suitable silica fibre of 204 μm diameter [20]. To be effective waveguides, fibres must be handled with care. Cleaving of the fibre must leave optically flat ends and both tension and constriction must be avoided.

Refracting telescopes

Large refractors are no longer built for astronomical observations because of the large, expensive domes required to house them, the difficulties of supporting large lenses at their edge, and the limited spectral range free from chromatic aberration. Nonetheless, a refractor can give a wider flat field than a reflector at a given focal ratio for a limited bandpass. Consequently, refractors of modest aperture are still built for astrometric work (as well as for planetary and solar patrols).

Figure 3.43 shows the elements of the 23 cm objective of the Hamburg Observatory astrograph [112]. The objective is an unsymmetrical

Fig. 3.42 Measured values of output focal ratio as a function of input focal ratio for a 204 μm diameter fibre. The dotted line corresponds to an f/∞ input. (Reproduced from [20].)

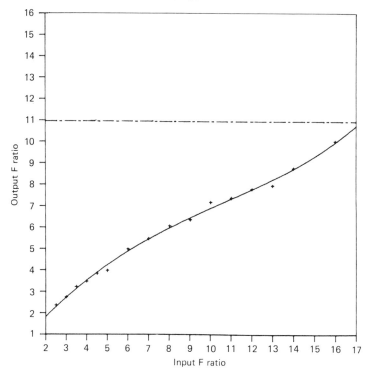

quintuplet with a focal length of 2.06 m which has a useful spectral range of 530 to 580 nm corrected for chromatic aberration. The plate scale is 100.47 arcsec mm^{-1} (i.e. 10 μm corresponds to 1 arc second). A field of $6° \times 6°$ can be taken on a plate of 215×215 mm.

The guiding telescope (focal length 2580 mm, aperture 190 mm) is rigidly attached together with the main telescope to a common central tube structure.

Telescope orientation

The faintness of astronomical sources means that telescopes in category (iv) must be able to track a source at the sidereal rate to compensate for the Earth's rotation for an exposure lasting between a fraction of a second and up to a day or more. The telescope pointing stability must be at least an order of magnitude better than the angular FWHM of the best point source images achieved by the telescope and detector combination.

The design goal of the Hubble Space Telescope is an r.m.s. pointing stability of 0.007 arcsec during a 10-hour exposure. This will be achieved using gyroscopes and the fine guidance system. Such high precision tracking cannot yet be achieved from the ground except, to some extent, with the use of active optics as discussed in Chapter 4. Gravitational stress, bearing roughness and atmospheric (or, for radio, ionospheric) effects introduce the principal limitations.

Precession and stellar coordinates are discussed in Chapter 2. Telescopes are mounted either altitude-azimuthly as is the Effelsberg

Fig. 3.43 The five elements of the 23 cm objective of the Hamburg Observatory astrograph. It has a focal length of 2.06 m, and a flat field of 10° diameter over the spectral range 490 nm to 590 nm. Published with the permission of Carl Zeiss, Oberkochen.

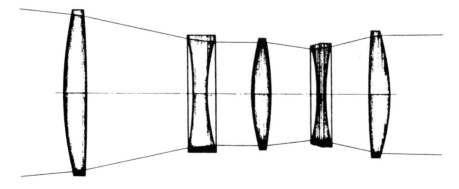

100 m telescope in Figure 3.17 or equatorially such as the telescope in Figure 3.41.

An equatorially mounted telescope rotates about the north-south axis (the polar axis) at a rate which just compensates for the Earth's rotation. The orthogonal declination axis allows the telescope to point to any part of the sky. The alt-azimuth mounting is normally adopted for larger instruments and has horizontal and vertical axes with variable drive rates to provide smooth tracking of the source. For wide field observations with an alt-azimuth telescope the detector must also rotate to compensate for field rotation.

An altitude-azimuth mounting is more compact for large optical telescopes than an equatorial mounting and the loss of accurate tracking through a small region near the zenith is offset by the availability of two quasi-stationary Gregorian or broken Cassegrain foci. These advantages are exploited in the Multi-Mirror Telescope (MMT) on Mount Hopkins of which a photograph is shown in Figure 3.44 [75]. Future large optical telescopes are likely to be built on altitude-azimuth mounts.

Atmospheric refraction

Atmospheric (or ionospheric in the case of long wave radio) refraction alters the apparent position of a source and also causes it to move more slowly than the reflex of the Earth's rotation when the source is setting, and faster when it is rising. Atmospheric refraction, R, is defined as [5]

$$R = z_t - z_a \tag{3.23}$$

where z_t and z_a are the true and apparent (refracted) zenith distances, respectively.

Figure 3.45 shows z_t plotted against R for a pressure of one atmosphere and a temperature of $10\,°C$ [5]. To determine R for other values of temperature (T in $°C$) and pressure (P in mmHg) multiply the value of R from Figure 3.45 by

$$P/(760(0.962 + 0.0038T)) \tag{3.24}$$

The variation of refractive index with optical wavelengths is discussed in Chapter 4. The variation of differential refraction with zenith angle at Mauna Kea is shown for a range of optical wavelengths in Figure 3.46 [380]. For on-axis spectroscopy it is possible to compensate for atmospheric dispersion with a variable dispersion prism. At high radio frequencies the refraction is approximately twice the optical value but largely independent of wavelength.

The differential effect of atmospheric refraction is significant across a wide field. For example [189], for the 6° field of the UK Schmidt Telescope described earlier in this chapter, the field is compressed in the vertical direction by 11 arcsec at a zenith distance of 45°. Even at the zenith the compression is about 5 arcsec in all directions.

For such a telescope, atmospheric refraction introduces a significant field rotation during an exposure. This effect can be minimised by changing the elevation angle of the polar axis according to the formula [409]

$$ME = \pm \arcsin((n-1)\cos\phi \cdot \sin\Delta/\cos(\phi-\Delta)) \tag{3.25}$$

where ME is the polar axis elevation above the (unrefracted) pole, Δ is the declination of the field centre, and ϕ is the latitude of the observatory.

Fig. 3.44 A view of the Multi-Mirror Telescope on Mount Hopkins, Arizona. There is a cluster of six 1.8 m $f/2.7$ primary mirrors. Each telescope directs a $f/32$ beam to the beam combiner such that the envelope of the six combined beams is $f/9$. MMT Observatory photograph.

The effect of a temperature gradient causing a 'bending' of the incoming starlight is discussed in Chapter 4. Ionospheric refraction is also discussed in Chapter 4.

Optical apodisation and the tapering of radio antennae [215]
Optical apodisation

The resolving power of a telescope is normally expressed in terms of the Rayleigh criterion (see equation (2.1)). This criterion, however, is only appropriate for the resolution of two close point sources of similar intensity [36]. A frequent problem in astronomy involves the mapping of faint structure, or a search for faint companions close to a bright source. For large telescopes above the Earth's atmosphere or for smaller ground based telescopes it can be useful to modify the point spread function or telescope transfer function.

In the classical diffraction pattern of a clear, circular telescope aperture the average intensity falls off as the inverse cube of angular distance from

Fig. 3.45 The variation of atmospheric refraction in seconds of arc, R, with the true zenith distance, z_t, at a pressure of one atmosphere and a temperature of 10°C.

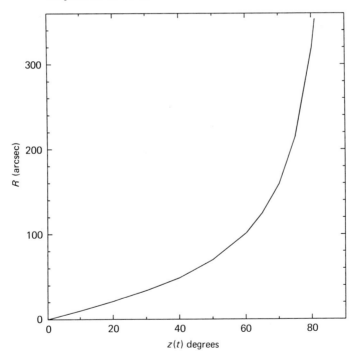

the central maximum. Relative to the intensity of the central maximum the first diffraction ring has an intensity of 1.7 per cent and contains 8.6 per cent of the energy while, for the tenth ring, the percentages are 10^{-3} and 0.2, respectively. The effects of atmospheric seeing and mirror surface imperfections are discussed in Chapter 4.

In most reflecting telescopes the secondary mirror or prime focus assembly together with its support spider causes a central obstruction which has the effect of increasing the energy in the diffraction rings at the expense of the central maximum. Several examples of diffraction patterns are shown in Figure 3.47 for different amounts of central obscuration including the case of a clear aperture. The ratio of the secondary shadow diameter to the diameter of the primary is given by d/a. For the case of $d/a = 0.3$, the intensity of the first diffraction ring becomes 4.4 per cent and the total energy 22 per cent which is a large change from the clear aperture pattern even though the secondary obstructs only 9 per cent of the incoming light [215].

Apodisation (removal of the 'foot' of the diffraction pattern) reduces the energy in the rings of the diffraction pattern by a modification of the shape or the transmission across the telescope aperture. This usually occurs at the expense of some broadening of the central maximum or Airy disc which reduces resolution by the classical Rayleigh criterion, and by an overall reduction in intensity. The simplest apodising mask is a radial, neutral density filter whose transmission varies smoothly from 100 per cent at the centre to zero at the edge of the mirror. An example is shown in Figure 3.48 for an apodiser providing maximum transmission [215]. The transmission function is shown by the dot–dash line, the clear aperture

Fig. 3.46 The difference (true–apparent) zenith angle, z, caused by atmospheric refraction and normalised by $\tan z$, is shown as a function of wavelength for Mauna Kea. (Published with permission from [79].)

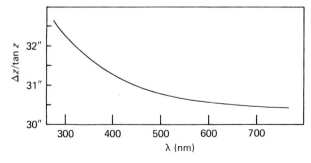

diffraction pattern by the dotted line, and the apodised pattern by the solid line.

The inverse of apodisation is also possible in which 'super-resolution' is achieved by narrowing the central maximum while enhancing the energy in the diffraction rings. This occurs naturally for telescopes with a large central obstruction as shown by the diffraction patterns in Figure 3.48.

Optical apodisation techniques have, so far, only been used for particular observational problems such as the 'detection' of Sirius B. In that case light in the point spread function was redistributed using a hexagonal mask over the telescope objective. For the proper orientation of the mask, contamination of the image of B by the nearby (10 arcsec) brighter (10^4) image of Sirius A was minimised [254].

Tapering of radio antennae and feed horns [60]

The angular resolution of available single radio telescopes is orders of magnitude worse than that available with single optical

Fig. 3.47 Examples of four diffraction patterns for telescopes with differing amounts of central obscuration, where d is the diameter of the secondary shadow and a the diameter of the primary. (i) $d/a=0$ (clear aperture), (ii) $d/a=0.1$, (iii) $d/a=0.2$, (iv) $d/a=0.3$. (Reproduced from [215].)

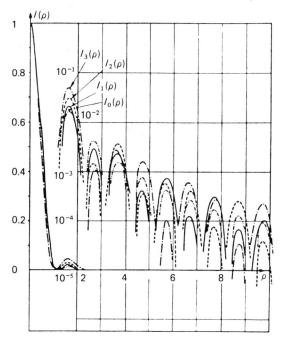

telescopes. For example, the diameter of the Airy disc (defined by the first minimum in the diffraction pattern) for the 100 m Effelsberg telescope operating at 3 cm is about a minute of arc compared with fractions of a second of arc for the larger optical telescopes (neglecting the effects of seeing discussed in Chapter 4). The higher order diffraction rings or side lobes corresponding to a point source include background radiation from a considerable area of the sky.

In heterodyne detection, radiation from the source must be delivered, in phase, to the receiver before it is mixed with the signal from the local oscillator. At radio frequencies this is achieved by mounting an appropriately shaped waveguide or feed horn at the telescope focus [23, 91, 299, 364].

For a single dish, the side lobes are suppressed by making the feed horn only just large enough to admit the Airy disc. This is equivalent to applying a weighting function similar to that in Figure 3.48 which emphasises the centre of the reflector.

Fig. 3.48 Example of an apodisation mask which also gives maximum transmission. Curve *a* is the pupil function, the solid line is the single mode response of the telescope, and the dotted curve is the clear aperture response. (Reproduced from [215].)

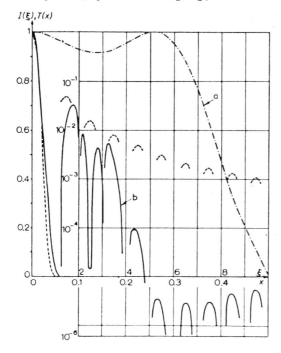

At the prime focus the weighting function must also prevent radiation from the ground from entering the feedhorn, as a result the weighting function must go to zero at the edge of the aperture.

Cassegrain systems allow more flexibility in producing tapering functions which use a larger proportion of the radiation falling onto the aperture [91]. By departing from strictly parabolic and hyperbolic surfaces a more uniform weighting can be achieved. The technique is known as beam-shaping. Further, since the horn looks away from the ground in a Cassegrain system the problem of parasitic ground radiation is removed.

When radio dishes are used in dilute apertures of the LBI type (see Chapter 6) the presence of side lobes is less objectionable. Consequently, the sensitivity of telescopes used in aperture synthesis can be significantly improved by using coaxial ring feeds which accept radiation in the first diffraction ring as well as the Airy disc [91]. The weighting function for such a coaxial ring feed on a Cassegrain telescope is compared with the more conventional prime focus weighting function in Figure 3.49.

An example of the suppression of unwanted side lobes in the case of array telescopes has already been given by the weighting or excitation function shown in Figure 3.13 for the 22 MHz array.

Fig. 3.49 Comparison of the maximum theoretical and observed efficiencies of parabolic antennae using a circular feed or coaxial feeds accepting radiation from one or more rings of the diffraction pattern. (Published with permission from [91].)

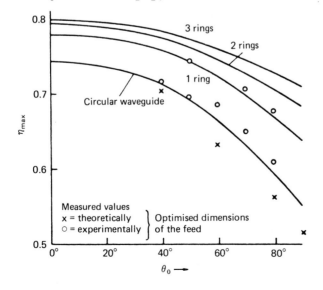

4

Seeing, speckles, and scintillation

Modulation (MTF) and optical transfer functions (OTF)

If the angular distribution of intensity in a source is a one-dimensional sinusoid of the form

$$f(\alpha) = a + b\cos(2\pi\phi\alpha + \Delta) \qquad (4.1)$$

where ϕ is the angular frequency, Δ is the phase, a is a pedestal or constant background, and b is the modulation of the sinusoid, as shown in Figure 4.1, then its image as formed by an optical system will also be a sine wave but with a different modulation [110].

The ratio of the intensity modulation in the image to that in the source is a function of angular frequency and it is called the modulation transfer function or MTF. The MTF is equal to the modulus of the Fourier transform of the spread function where the spread function is the image formed by the optical system of a line source. The optical transfer function or OTF is the Fourier transform of the line spread function.

The MTF is normalised to unity at zero angular frequency and, if the line spread function is both real and symmetric, the optical transfer function and the modulation transfer function are identical.

Fig. 4.1 A one-dimensional sine wave of spatial frequency ω, modulation b/a, and phase Δ.

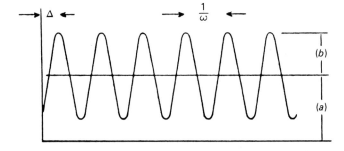

Telescope transfer function and atmospheric transfer function

The value of Fourier techniques lies in the convolution theorem which states that the convolution of two functions is equal to the product of their Fourier transforms. Wavefronts entering the telescope are subject to many perturbations all of which contribute ultimately to the image quality. In the case of optical observations the OTF is the product of the telescope and atmospheric transfer functions. It is also more appropriate to talk in terms of the point spread function which is the two-dimensional analogue of the spread function.

The telescope transfer function can be expected to remain fairly constant during an exposure. Figure 4.2 shows the on-axis MTF at wavelengths between 100 nm and 400 nm for the primary of the Space Telescope. The mirror gives a diffraction-limited performance in the visible. The r.m.s. surface deviation from the theoretical figure for the Hubble Space Telescope primary is 1.3×10^{-8} m (i.e. $\lambda/50$ at 632.8 nm).

Fig. 4.2 The on-axis Modulation Transfer Function (MTF) for the Hubble Space Telescope at 100, 200, 300, and 400 nm. The mirror has been figured to give diffraction limited performance at 633 nm with a surface deviation of $< \lambda/50$.

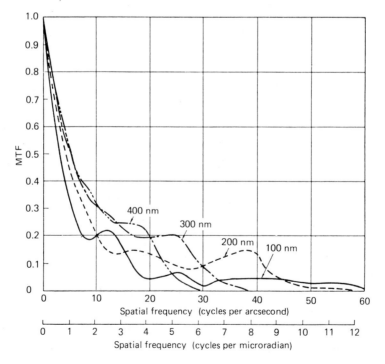

The increasing power in side lobes in the ultraviolet arises from increasing wavefront errors at shorter wavelengths.

The atmospheric transfer function can be determined directly by measuring the spatial coherence of the radiation entering the telescope from a coherent source (a star). If the telescope beam is displaced or folded and recombined with the original beam the spatial extent of fringes in the resulting interferogram will indicate the degree of coherence in the original beam [111, 330]. Figure 4.3 shows an example of interference fringes achieved by folding one half of the entrance pupil of a 0.9 m telescope on itself under a variety of atmospheric conditions [109]. The optical arrangement is shown in Figure 4.4. The extent of the fringes indicates that the incident wavefront was correlated over a few centimetres. Such a result allows a direct calculation of the atmospheric transfer function which, in turn, is the Fourier transform of the seeing profile. The physics of 'seeing' is discussed in detail below.

Optimum sampling [110, 404]

In the process of detection, the individual sensing elements of the detector sample the telescope image (as distorted by the atmosphere and tracking errors). The spatial frequency spectrum in the image finally constructed from the detector output depends on the frequency content of the image prior to detection and the MTF of the detector. The latter critically depends on the spacing between, and the sensitivity variations within, the individual sensing elements.

If the linear spacing of the sensor elements, δx, is uniform, then the Nyquist or folding frequency, ω_{max}, is given by $(2\delta x)^{-1}$. The sampling theorem states that, if a function of x is defined in the range of $\Delta x = m$, then its Fourier transform is completely described by values at frequency intervals of $1/m$. Most point, or line, spread functions tend to be Gaussian in form and, if $(2\delta x)^{-1}$ is not to become very large, a compromise in sampling interval must be used. As a rule of thumb if δx is one-quarter of the half-width at half height of the line spread function the absolute error in the computed MTF should be less than 0.005. In other words such a sampling interval would recover virtually all of the frequency information in the original image.

Astronomical images are usually undersampled. This leads to the phenomenon of *aliasing* which is illustrated for the simplest of cases in Figure 4.5. A continuous sinusoid of 40 cycles mm^{-1} and another of 10 cycles mm^{-1} are sampled by a detector with a pitch of 20 μm. From the

Fig. 4.3 Photographs of fringes obtained when observing a bright star in (a) poor, (b) average, and (c) good seeing, by folding one half of the entrance pupil of a 0.9 m telescope on itself using the shearing interferometer arrangement shown in Figure 4.4. (Published with permission from [109].)

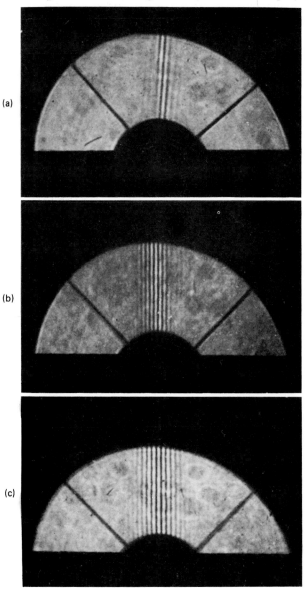

(a)

(b)

(c)

point of view of the detector the two frequencies are indistinguishable. In general the alias frequencies associated with the Nyquist frequency, ω_{max}, and a spatial frequency in the image, ω, are

$$(2\omega_{max} - \omega), \qquad (2\omega_{max} + \omega), \qquad (4\omega_{max} - \omega), \qquad (4\omega_{max} + \omega),$$

etc.

Fig. 4.4 The shearing interferometer used to obtain the fringes shown in Figure 4.3. (Published with permission from [111].)

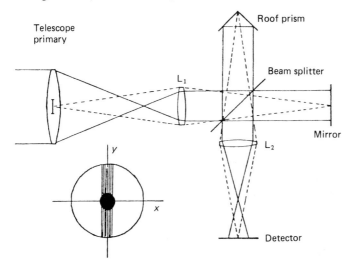

Fig. 4.5 An example of aliasing. Continuous sinusoids of 40 and 10 cycles mm^{-1} are sampled by a detector with individual sensing elements on $20\,\mu m$ centres (50 mm^{-1}). $\omega_{max} = 25$ cycles mm^{-1} and $\omega = 10$ is confused with $2\omega_{max} - \omega = 40$.

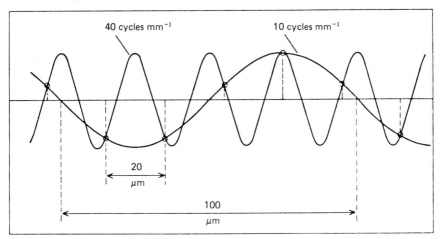

This means that structure within the image of a higher frequency than ω_{max} will appear at lower beat frequencies in the constructed image, i.e. each of the odd terms above. Where aliasing is significant this has the effect of increasing the signal noise at low frequencies.

A useful limit for significant aliasing is $\sigma \geqslant \delta x$ [404], where 2σ is the width of the point or line spread function at an intensity of $1/e$ times the central intensity.

Atmospheric turbulence

In Chapter 1 it was demonstrated that most stars have angular diameters which are much smaller than the Rayleigh limit of available optical telescopes. The 'image' of a star represents the response of the telescope to a coherent point source after the light has passed through the Earth's atmosphere.

The air has a variable refractive index, n, with a mean value close to 1 and a variable portion of the order of 10^{-6} or less. At 500 nm the value of n is approximately 1.0003 at sea level and 1.0001 at 10 km altitude [323]. A refractive modulus, N, is defined as

$$N = (n-1) \cdot 10^6 \tag{4.2}$$

At optical wavelengths the approximate expression (Cauchy formula) for N at STP is

$$N = 272.643 + 1.2288/(\lambda^2 \cdot 10^{-6}) + 0.035\,55/(\lambda^4 \cdot 10^{-12}) \tag{4.3}$$

where λ is in nm.

The variation of N with atmospheric pressure, P (Pa), and temperature, T (K), is given approximately by

$$N = 0.79 P/T \tag{4.4}$$

Variations in water vapour content have no significant effect on the refractive index at optical wavelengths. From equation (4.4) it is possible to find an approximate expression for the fluctuating part of the refractive modulus, ΔN, as a function of the fluctuations in temperature, ΔT, from the equation of state for adiabatic conditions in a perfect gas,

$$\Delta N = 2\Delta T \cdot (P/T^2) \tag{4.5}$$

This assumes that small-scale, turbulent atmospheric motion is adiabatic.

The atmospheric turbulence discussed in this section has several origins. The principal one is convection. Air heated by conduction at the Earth's surface rises into cooler air, expands, and continues to rise with the cooler air descending towards the surface. High winds, particularly those associated with the jet stream, generate turbulence through wind shear. If an observatory is in the lee of high mountains then these can introduce large scale disturbances.

Energy is deposited into turbulence at a so-called outer scale, L_0, of the order of 100 m. This cascades to smaller and smaller scales until viscous dissipation slows the turbulent motions sufficiently that heat conduction smoothes out any further temperature fluctuations at a scale of a few millimetres.

The Kolmogorov spectrum is the one most frequently used to describe isotropic atmospheric turbulence. It is characterised by a radial power spectrum of the form

$$G(k) = \text{constant} \cdot (kL_0/2\pi)^2/(1 + (kL_0/2\pi)^2)^{11/6} \qquad (4.6)$$

where k is the wave number [36]. This equation is only appropriate for thermally generated turbulence and does not apply to the larger scale image motion induced, for example, by mountains.

Speckles [30, 233]

Wavefronts arriving above the Earth's atmosphere from a star are assumed to be plane-parallel. The temperature inhomogeneities caused by atmospheric turbulence produce small scale variations in the refractive index according to the relation in equation (4.5). These variations produce corrugations of up to several wavelengths' amplitude in the wavefronts before they reach the telescope. As a result the telescope optics cannot form the classical diffraction pattern (the Airy disc) appropriate to a single point source.

Those parts of the wavefront which are in phase with each other form fringes at the telescope focus according to the geometry shown in Figure 4.6. Two points in phase which are a distance d apart at the telescope

Fig. 4.6 The basic geometry involved in the formation of speckles in the focal plane of a telescope.

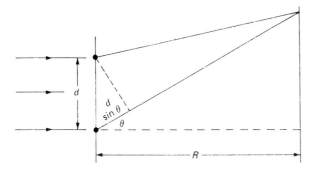

mirror will form fringes, or speckles as they are better known, at angular separations of

$$\theta = n\lambda/d \quad \text{rad} \tag{4.7}$$

where n can be any integer including 0. If the bandpass of the radiation is restricted to $\Delta\lambda$ then, for the highest order detectable speckle,

$$n_{max} = \lambda/\Delta\lambda \tag{4.8}$$

A turbulence coherence length, r_0, can be defined which is a function of the power spectrum of the turbulence, the refractive index, the wavelength of the transmitted starlight, and the air mass above the telescope. Within r_0 the r.m.s. phase fluctuation of the electric vector across the wavefront is one radian. In homogeneous turbulence

$$r_0(\lambda) = r_0(\lambda_0)(\lambda/\lambda_0)^{1.2}(\cos z)^{0.6} \tag{4.9}$$

where z is the zenith angle made by the telescope beam with the vertical. The turbulence coherence length can be measured directly from the width of the fringe pattern in Figure 4.3.

From a full analysis, the principal deformation at the scale of r_0 consists of a tilting of the wavefront. From the definition of r_0 and relations (4.7) and (4.8) it follows that:

(a) the angular spacing of the closest speckles is proportional to λ/D where D is the telescope diameter,

(b) the envelope of all the first order speckles is defined by a circle of angular diameter λ/r_0 (it also contains higher order speckles from pairs of in-phase points more widely spaced than r_0),

(c) following from (a) and (b), the number N, of speckles is given, approximately, by

$$N = (\lambda/r_0)^2/(2\lambda/D)^2 = D^2/(2r_0)^2 \tag{4.10}$$

because each pair of in-phase points would tend to produce only a single detectable speckle (since wavefront tilt is the dominant deformation),

(d) if the source is an extended object of angular size ϕ, contrast in the speckles begins to deteriorate when ϕ is of order $2\lambda/D$.

Speckle interferometry

The angular distribution of intensity in a speckle pattern is equivalent to a Fourier transform of the incoming wavefront convolved with angular structure in the source and the telescope point spread function in the absence of the atmosphere [240, 253, 355]. Provided exposure times are $< t_0$, the turbulence coherence time (described below in section on seeing), it is possible to use the information from a series of specklegrams to separate the effect of atmospheric seeing (atmospheric transfer function) from the telescope response to structure in the source. The technique, known as speckle interferometry, is used regularly in measuring the elements of close binary stars.

Figure 4.7 shows speckle patterns of the three bright stars, Vega (α Lyrae), Betelgeuse (α Orionis), and Capella (α Aurigae) obtained by Labeyrie and his collaborators with the Hale 5 m telescope [239]. The exposure times were 0.01 s, the wavelength 500 nm, and the bandpass $\Delta\lambda = 20$ nm. The patterns are typical of three different types of stellar source:

   (i) Vega is a single star whose angular diameter is much less than the telescope Rayleigh limit (see discussion in Chapter 2),

  (ii) Betelgeuse is a single star whose angular diameter is greater than the telescope Rayleigh limit,

Fig. 4.7 Direct, highly magnified photographs of Betelgeuse, Capella, and Vega taken with the 5 m Hale telescope at Mt Palomar. The exposure times were 0.01 s, the bandpass was 20 nm, at a wavelength of 500 nm, and the f/ratio was 200. Each image shows a different speckle structure. The three lower images are Fourier transforms generated when the individual specklegrams act as diffraction screens in a collimated laser beam. The intense zero orders have been suppressed by an occulting spot and the individual images are accumulations from many specklegrams. (Published with permission from [239].)

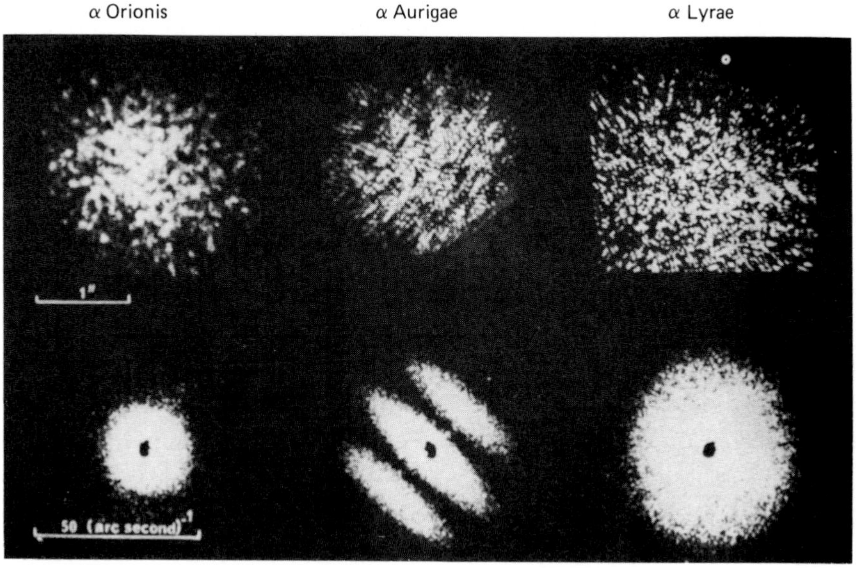

α Orionis α Aurigae α Lyrae

(iii) Capella is a close binary star with an angular separation which is larger than the telescope Rayleigh limit but whose component stars have angular diameters which are less than the Rayleigh limit.

Each specklegram displays quite a different character. There are some 500 speckles in the snapshot of Vega. From equation (4.10) this leads to a value of 11 cm for r_0, which is typical for a good observatory site.

The speckles in the case of Betelgeuse are more diffuse than those for Vega while the speckles for Capella appear double but individually as sharp as those for Vega.

Specklegrams were taken of each star at intervals $\gg t_0$. Each one was used as a diffraction screen in a collimated laser beam and the transmitted light was focussed and photographed. The photographs from the series were added to give the results shown in Figure 4.7. The angular intensity distribution in each photograph is a measure of the squared modulus of the Fourier transform of the spatial frequencies present in the specklegrams. The strong zero order has been suppressed by the dark spot at the centre of each photograph.

The Fourier transforms can be interpreted as follows.

(i) That for Vega gives the angular frequency response of the telescope to a point source.

(ii) That for Betelgeuse shows the high frequency cutoff caused by the finite angular diameter (about 0.06 arcsec) of the star.

(iii) Capella shows a classical two-slit diffraction pattern. The orientation indicates the direction of the line of centres of the two component stars in the binary. The spacing of the pattern is inversely proportional to their angular separation (about 0.06 arcsec). Since the envelope of the whole pattern is equal to that for the unresolved star Vega, the discs of the individual stars must be smaller than the telescope Rayleigh limit.

Searches for, and measurements of, close binary systems are undertaken routinely using the above technique [120, 239]. Figure 4.8 shows schematically the observational arrangement used on the 4 m Mayall telescope at KPNO while Figure 4.9 shows a schematic at the optical processor [278]. The observations are recorded on Tri-X film through a 20 nm wide filter at a central wavelength of 550 nm. With this combination individual speckles have diameters of about 0.05 arcsec for a point source. Figure 4.10 shows a composite transform made by McAlister from 50 specklegrams of the triple star η Orionis. The fine set of fringes come from the previously known binary, A–B, of 1.5 arcsec separation; the broad fringes show that A is a previously unresolved binary of 0.044 arcsec separation at the time of observation in 1975.

This analogue treatment demonstrates the essentials of the speckle interferometric technique. Basically, the time-averaged autocorrelation function must be computed for the angular intensity distribution in the stellar image for exposures short compared with t_0. With the advent of area detection systems which register the position of each detected photon the technique can be extended digitally to much fainter stars than in the analogue technique. The autocorrelation function can be calculated either off-line or in real-time [59, 191, 193, 194].

Rapid optical switching arrangements can be used to give nearly monochromatic simultaneous speckle patterns associated with neighbouring wavelengths. From these, an image of the star or its chromosphere, for example, can be reconstructed when one wavelength is in the continuum and the other in an emission line [27].

Fig. 4.8 A schematic of the Kitt Peak National Observatory speckle camera used at the Cassegrain focus of the 4 m telescope. The image tube gain is $> 10^5$. The Risley prisms compensate for atmospheric dispersion. The microscope objective gives an effective focal length of 476 m for the telescope. (Published with permission from [278].)

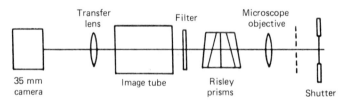

Fig. 4.9 Schematic of the optics which produce composite Fourier transforms of the specklegrams obtained with the camera in Figure 4.8. The liquid, having a similar refractive index to the film, eliminates the effect of scratches on the film. The Ronchi grating can be used to calibrate position angle measurements. (Published with permission from [278].)

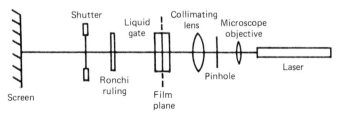

The faint limiting magnitude of the speckle autocorrelation technique is reached when photons from the sky and the detector noise contribute a significant fraction of the photon events detected within the stellar image.

Amplitude interferometry [108, 331]

A visibility (see Chapter 2) can be derived for the central fringes obtained with the shearing interferometer in Figure 4.4. Two images of the telescope exit pupil are superimposed, with one rotated relative to the other. The maximum baseline, and hence angular resolution, is obtained for a rotation of 180°. By varying the angle of rotation the fringe visibility at different angular frequencies is sampled in two dimensions. This is the basis of the amplitude interferometric technique.

Amplitude interferometry is limited in just the same way as speckle interferometry, by the wavelength dependence of atmospheric turbulent distortion, to bandwidths of a few tens of nanometres. The fringe visibility is insensitive to atmospheric turbulence and telescope aberrations, but exposures must be $<t_0$, the turbulence coherence time (discussed in the

Fig. 4.10 The composite Fourier transform of η Orionis from 50 specklegrams taken with the camera shown in Figure 4.8. The fine set of fringes come from the previously known binary of 1.5 arcsec separation. The broad fringes indicate that component A is a previously unresolved binary of 0.044 arcsec separation at the time these observations were taken in 1975. (Published with permission from [278].)

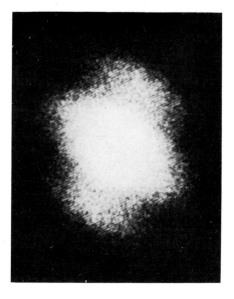

section on seeing below). On the other hand the interferograms are free from speckle noise.

In theory, an observer can estimate by eye directly the separations of binaries and stellar diameters. For example, for a sufficiently extended star such as Betelgeuse, the stellar disc is resolved when the fringes disappear at a particular value of the rotation [331]. Scintillation (see section below) is an important source of error in estimating fringe visibility. Consequently, visibility must be estimated from an average of several interferograms.

Seeing [370]

The broadening of stellar images by atmospheric turbulence is called 'seeing' (sometimes 'astronomical seeing'). Speckle patterns change too rapidly for the eye to follow and they are poorly defined in white light. Consequently, the integrated effect of the longer exposure is closer to the image registered by the eye. At a really good site the visible seeing, α, is typically about one arc second or less and corresponds roughly to the envelope diameter of the first order speckles defined above in (a) as

$$\alpha = 2 \times 10^5 \cdot \lambda/r_0 \quad \text{arcsec} \tag{4.11}$$

Equation (4.11) implies that the angular size of an image at a given wavelength is independent of the telescope diameter when $D > r_0$. For the 5 m telescope the optical image quality on long exposures is little better than the combined images from an array of 2000 11 cm telescopes, i.e. with diameters $= r_0$.

From relations (4.9) and (4.11),

$$\alpha(\lambda) = \alpha(\lambda_0)(\lambda_0/\lambda)^{0.2}(\sec z)^{0.6} \tag{4.12}$$

which means that the size of the seeing disc decreases towards longer wavelengths [355, 386]. This appears to be confirmed by the observations in Figure 4.11 [58]. Seeing was estimated simultaneously at 550 nm and at 10 μm on a 1.5 m telescope. The seeing was both poor and variable. The estimates at both wavelengths are shown together with the predicted correlation from equation (4.12).

The diameter, θ, of the Airy disc is

$$\theta = 2 \times 10^5 (2.44\lambda/D) \quad \text{arcsec} \tag{4.13}$$

Consequently, for a 4 m telescope the zenith seeing ($z = 0$) and diffraction limited images are the same size, 0.4 arcsec, at 14 μm when the zenith seeing is one arc second at 500 nm. This means that the best ground based images will be obtained in the infrared with large telescopes.

In a telescope with an aperture smaller than or equal to r_0, instead of a multitude of speckles there will be a single diffraction-limited image moving about on the time scale with which turbulent cells pass through the telescope beam. In fact, this effect is used in the testing of potential sites for large optical telescopes because the envelope of stellar image motion in a small telescope should, to a first order, define the seeing disc in a larger instrument.

A turbulence coherence time, t_0, can be defined as the time scale on which the turbulent pattern is sufficiently stationary that the speckle pattern is not significantly blurred during an exposure. If $r_0 \ll D$ then t_0 varies as r_0/V, or as D/V when r_0 is of order D, and as $(r_0/V)^{0.5}$ for

Fig. 4.11 Comparison of the estimated simultaneous seeing disc diameters at 550 nm and at 10 μm using a 1.5 m telescope. (Reproduced from [58].)

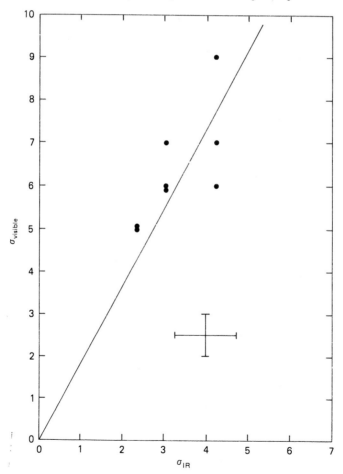

intermediate values (where V is the wind velocity). Figure 4.12 shows t_0 values for the intermediate case from infrared observations made with a 1.5 m telescope on Tenerife [355]. Again, t_0 like r_0 increases with λ.

The limiting diameter of any telescope to be used for heterodyne detection is r_0, otherwise phase distortions in the wavefront will mean that the incoming radiation mixes with the radiation from the local oscillator over too wide a range in phase. The dependence of r_0 on $\lambda^{1.2}$ means that this limitation is particularly severe at optical wavelengths.

Scintillation [144, 291, 323]

When the entrance pupil of a telescope is examined in the light of a bright star a pattern of bright and dark waves or shadows can be seen moving across it. The contrast between bright and dark areas is typically ten per cent and is caused by the 'focussing' effect of wavefront corrugations. Curvature concentrates more or less energy into different areas of the wavefront depending on the sign of the curvature and the height of the turbulent cells in the Earth's atmosphere. As the shadow pattern is blown across the telescope beam it generates the rapid fluctuations in the apparent brightness of a star which are known as 'scintillation'. Scintillation causes the twinkling of stars seen with the unaided eye which, because it has a small aperture, tends to sample only single turbulent cells. This effect tends to be averaged out as the telescope aperture increases.

A scintillation index can be defined as the variance in the radiation fluctuations observed for a bright star (i.e. where the photon shot noise is negligible). The character of the index varies depending on the relative

Fig. 4.12 The observed variation of the turbulence coherence time, t_0, with wavelength. The solid line is approximately a $\lambda^{0.5}$ law. (Published with permission from [355].)

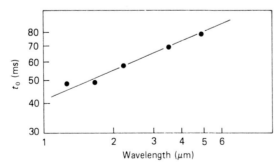

sizes of r_0 and the telescope aperture. In general two scintillation characters can be identified depending on the relative sizes of r_0 and D:

(i) If D and r_0 are of the same order of size the scintillation index will be around 20 per cent on an average night. The index decreases with the $-\frac{7}{6}$ power of the wavelength thereby becoming negligible in the infrared. The contribution of a turbulence layer to the scintillation increases as the $\frac{5}{6}$ power of the height above the telescope. In fact, the scintillation is produced in two principal regions, some 25 per cent between 1 and 2 km, and 75 per cent in the tropopause and lower stratosphere (between 8 and 15 km).

(ii) When $D \gg r_0$ the effect of the lower level of turbulence is heavily filtered out and the principal contribution comes from the 8 to 15 km level. The contribution of turbulent layers increases as the height squared as opposed to the $\frac{5}{6}$ power and decreases as the $-\frac{7}{3}$ power of the aperture, D.

Most theoretical studies of scintillation are based on ray optics, which is valid only so long as the smallest turbulent elements can be treated as primitive lenses. If they are too far removed from the telescope (distances exceeding 20 km for turbulent eddies of 10 cm), diffraction effects will dominate. At large zenith angles the distances to the turbulent elements rapidly increase and a 'saturation' sets in with no further increase in the scintillation index.

Optical image profiles (long exposure OTF)

Astronomers often assume that the central few arc seconds of long exposure stellar images have a Gaussian distribution of radial intensity, but it has been pointed out that the intensity profile for a large telescope is given by the two-dimensional Fourier transform of the atmospheric transfer function $B(f)$, where f is the angular frequency (assuming that the telescope has a sufficiently good figure and a large enough aperture that one can ignore the telescope transfer function) [332]. Direct measurements show that

$$B(f) = \exp(-3.44(\lambda f/r_0)^{5/3}) \tag{4.14}$$

Because the power $\frac{5}{3}$ is close to 2, the profile is indeed almost Gaussian.

In Figure 4.13 theoretical profiles based on $B(f)$ and values of $r_0 = 6$, 9.3, and 12 cm are compared with an observed stellar profile shown by the points. The observations were obtained at V with the CFH 3.6 m telescope on a night of average seeing. The theoretical profile for $r_0 = 9.3$ cm gives good agreement out to at least 6 arcsec. At greater distances an α^{-2}

dependence often dominates because of the scattering from fine scratches, dust, and contaminants which are always present on a mirror surface [88, 228, 234, 432]. It was pointed out in Chapter 3 that diffracted intensity falls off as α^{-3}.

Choice of optical telescope sites

Observatory sites with good seeing (1.5 arcsec or less) normally lie above a temperature inversion by night. In an inversion the air temperature increases with altitude which prevents convective mixing of the air from below the level of the observatory into the air above it. This arises naturally at night under dry, clear, conditions because the ground cools off more rapidly than the atmosphere. Around sunset the atmosphere becomes isothermal and eventually a temperature inversion develops [183, 285].

When the telescope is at the highest point in the immediate surroundings, cold air will not drain over it as the inversion develops. In

Fig. 4.13 Theoretical, long exposure, stellar image profiles for three values of r_0 are compared with observations (solid points). The intensity scale is logarithmic. The observations were made in average seeing through a V filter with a CCD at the prime focus of the Canada–France–Hawaii 3.6 m telescope. Observations courtesy of P. Hickson, calculations by P. Bennett, U.B.C.

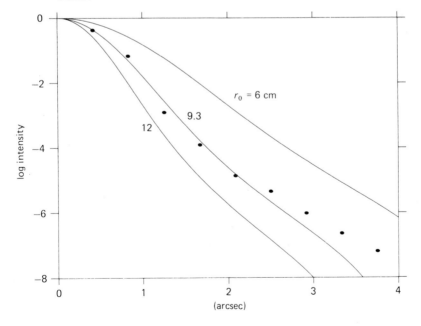

fact, the telescope is often deliberately placed some tens of metres above the ground in addition to being on a summit. The best observatory sites have a permanent inversion for most of the year, have laminar air flow in the prevailing wind, and lie outside the normal seasonal zones of the jet stream. To be classified as an excellent site there must also be little precipitable water vapour above it and the sky must be dark (points which are dealt with in more detail in Chapter 2).

These rather stringent criteria explain the success of the Californian and Chilean sites where the prevailing winds are from the west with no intervening mountains to perturb the laminar flow developed in thousands of kilometres' passage across the Pacific Ocean. Further, the coastal waters are cold, which maintains not only a low humidity but also the necessary inversion. Mauna Kea at nearly 4300 m is the highest mountain in the Hawaiian islands. The predominantly north-east trade winds also maintain a highly laminar air flow, and the combination of the altitude and the surrounding ocean appears to maintain an inversion.

The average surface brightness in the core of a stellar image depends upon M, the coefficient of surface brightness, which also provides a useful figure of merit for a telescope and its site,

$$M = (D/\alpha)^2 \tag{4.15}$$

where D is the diameter of the telescope primary and α is the angular diameter of the seeing disc. M indicates both the potential rate of energy collection and the degree of concentration of the image, both of which are critical in the telescope's ability to detect objects against the sky background.

To a first order the observed surface brightness of an unresolved source will be simply Mf, where f is the source irradiance (see Chapter 1). This observed quantity is not to be confused with the radiance, B, which is the actual surface brightness of the source. (Unfortunately α in normal convention is used for both seeing and the actual angular size of the source.)

The choice of a site usually has additional constraints such as accessibility and politics. An even distribution through the year of clear sky and night-time hours and a short astronomical twilight (defined as the period from sunset to the time that the Sun is 18° below the horizon, or the corresponding period before sunrise) are normally considered important. The variation of the number of potential dark hours per year is shown in Figure 4.14 as a function of geographic latitude [429]. This plot strongly favours sites within about 30° of the equator. Such sites also have a more even distribution of dark hours throughout the year. On the other hand observatories at high latitudes have seasons when circumpolar sources can be observed continuously for longer periods than at lower latitudes [204].

In order to maintain an inversion locally the effects of daytime heating and heat sources within the telescope dome must be minimised. Where

Fig. 4.14 The variation of the number of dark hours per year as a function of geographic latitude and acceptable twilight. (Reproduced from [429].)

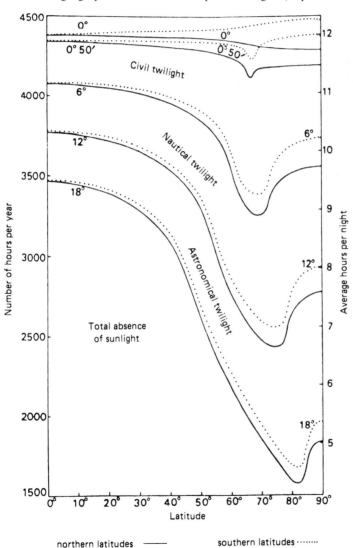

northern latitudes ——— southern latitudes ········

there is a temperature gradient of 1 K m^{-1} perpendicular to the telescope beam, the beam will be curved by 0.2 arcsec m^{-1} convex towards the side of higher temperature [9].

The dome is often painted with a titanium oxide paint to give it a high infrared albedo and therefore minimise solar heating. In some cases the observing floor underneath the telescope is refrigerated to less than the outside ambient temperature. With smaller telescopes and domes the night air is frequently drawn in by powerful exhaust fans to maintain the interior temperature at the same level as that outside. One menace of modern electronic equipment is large heat dissipation (often several kilowatts) which has to be ducted or neutralised at the telescope.

Seeing compensation techniques

There is considerable interest in techniques which sense and correct wavefront deformations either in real-time or in post-detection processing [181, 380]. Such techniques could recover the point source image of which the telescope is capable. Several schemes involving active optics have been used successfully to improve the images of single stars. Unfortunately, as atmospheric turbulence is at an effective height of several kilometres, wavefront errors are only correlated for points a few arc seconds apart. The solid angle over which wavefront errors are correlated is known as the isoplanatic patch.

Figure 4.15 shows the schematic of a real-time seeing-compensation system [180]. It is based on a Monolithic Piezoelectric Mirror (MPM). This device can be deformed in zones piezoelectrically by varying the voltages applied to the actuators. Referring to Figure 4.15, a fraction of the starlight from the MPM is diverted by a beam splitter to a wavefront shearing interferometer. The interferometer measures the local slope of the wavefront and provides corrections to the MPM.

The light beam within the interferometer is divided into X and Y branches which allow the tilt of the wavefront to be measured in orthogonal directions. Lenses L2x and L2y focus the stellar image formed by the telescope onto the two Ronchi gratings Gx and Gy. The gratings generate the wavefront shear through the 0, 1, and −1 order images of the surface of the MPM which are focussed onto the arrays of detectors Dx and Dy by the lenses L3x and L3y. The gratings rotate and have alternate opaque and clear radial lines. They intersect the X and Y beams at right angles and shear the beam in orthogonal directions. The rotation chops the speckle pattern and thereby the light to different pairs of detectors in the array, producing an alternating output. The phase of the alternating signal from each detector is compared with that from LED reference sources imaged on the gratings (X ref and Y ref). From this comparison it is possible to reconstruct the local tilts in the wavefront and to generate a correction signal to the appropriate actuators in the MPM.

The number and spacing of the detectors in each of Dx and Dy are the same as the number of actuators on the MPM which in turn is dictated by the relative sizes of D and r_0. Some 2000 independent actuators would be necessary to provide adequate real-time seeing compensation on the 5 m telescope at 500 nm according to the speckle pattern for Vega. If compensation is required outside the isoplanatic patch then, in theory, additional MPMs could be introduced for off-axis images.

The cost of such seeing-compensation systems for large telescopes is likely to remain prohibitive for some time. Actually, considerable image improvement is possible by use of a single, two-axis, tilting mirror (with the error signal being derived from a quadrant detector) because one of the major contributions to seeing at good sites is an overall tilt of the wavefront which introduces motion of the whole image. This is nicely

Fig. 4.15 Block diagram of a real-time seeing-compensation system. Wavefront errors are measured in the interferometer and corrections are applied through the deformable mirror. (Published with permission from [180].)

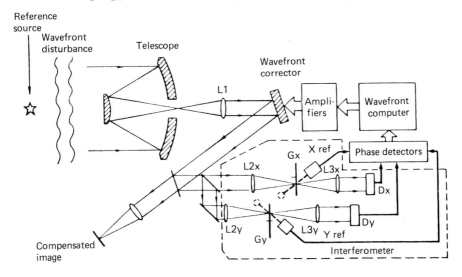

illustrated in Figure 4.16 which shows image profiles for the bright double star γ Pegasi with:

(a) no seeing compensation,

(b) overall tilt compensation,

(c) both tilt and MPM wavefront correction.

The observations were made with a 30 cm telescope and a 21-element MPM. The brighter, primary image of the binary was used to provide corrections. It is clear that both components are improved with the tilt correction. However, while the full wavefront correction gives a marginal additional improvement in the primary image, the secondary image has slightly deteriorated. This implies that the isoplanatic patch for overall tilt was greater than 4 arcsec radius while that for higher order compensation was less than 4 arcsec.

Scintillation at radio frequencies

The speckle phenomenon caused by atmospheric turbulence is confined to the optical and infrared regions. At shorter wavelengths the atmosphere is opaque and observations have to be made from rocket or satellite altitude. At radio frequencies there is a very similar phenomenon which is caused by electron scattering [95, 317, 337].

Electron density irregularities in the ionosphere, the solar wind and corona, and the interstellar medium distort the wavefronts of radio waves

Fig. 4.16 Image profiles for γ Pegasi obtained with the seeing compensation instrument shown in Figure 4.15 for: (i) no correction, (ii) overall wavefront tilt correction only, (iii) both tilt and detailed wavefront correction. (Published with permission from [180].)

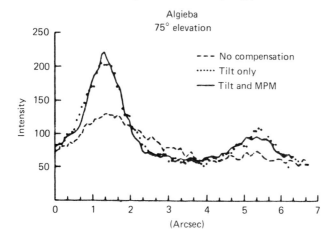

passing through them, thereby generating the equivalent of optical speckles and scintillation. The scale lengths of these irregularities which correspond roughly to r_0, the coherence length of atmospheric turbulence, are normally larger than the aperture of any single radio telescope. Consequently, while no speckles can be detected, scintillation, as with the twinkling of stars seen with the naked eye, is very important.

A scale size of between 500 m and a few kilometres has been measured for ionospheric irregularities either by following the 'shadow' pattern on the ground with spaced receivers or from the scintillation of small radio sources (either artificial satellites or compact cosmic sources) observed with a large radio telescope such as that at Arecibo [1]. The height of the ionosphere is between 300 and 500 km and image wander and scintillation fluctuations have a time scale of the order of half a minute. Any source with an angular diameter of less than some ten minutes of arc can be considered coherent for the geometry of the ionospheric irregularities and, in consequence, would show scintillations.

Interplanetary and ionospheric scintillations can be easily separated from each other because the former have a time scale of about one second and are only detected for sources with an angular size < 0.5 arcsec. Figure 4.17 illustrates the geometry involved in the scattering of waves from a distant radio source by the interplanetary medium. The radio source is effectively at infinity and scattering takes place on irregularities in the solar wind blowing across the line of sight [318]. If the source has an angular size $< \lambda/a$, where a is the scale length of the irregularities, then the radiation can be treated as coherent.

Unlike speckle interferometry, the limiting angular diameter estimated from radio scintillation is critically dependent on an accurate model for the electron density structure in the plasma [97, 98, 101, 344]. The theory of interplanetary scintillation (IPS) considers that the perturbations to the wavefront arise in a thin diffracting screen at a distance of around one AU from the telescope.

There are two measured quantities of importance. One, the scintillation index, is the r.m.s. fluctuation in the received signal. The other is the power spectrum of the signal fluctuations. The latter depends on the r.m.s. phase fluctuations generated in the screen and the quantity z/z_0. z is the effective distance to the screen, $z_0 = 2\pi a^2/\lambda$ (the Fresnel distance to an irregularity), a is the scale length of an irregularity in the screen, and λ is the wavelength of the detected radiation.

Some radio telescopes have been built specifically to exploit

interplanetary scintillation for the measurement of radio source
diameters. One of these, the 81.5 MHz array at Cambridge, already
described in Chapter 3, is celebrated for the serendipitous discovery of
pulsars by Jocelyn Bell [196]. The wavelength of 3.7 m was chosen
because scintillation effects are more pronounced at long wavelengths.

Fig. 4.17 The scattering geometry for radiation from a distant radio source
seen through the interplanetary medium. Inhomogeneities in the solar wind
are considered to be moving across the line of sight with velocity V. If the
angular size of the source is $< \lambda/a$, where a is the scale length of the
irregularities, then the radiation can be considered coherent. (Published with
permission from [318].)

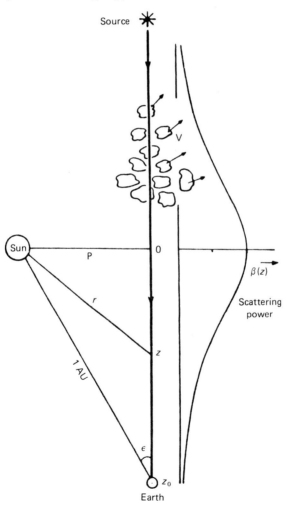

Figure 4.18 shows the variation of scintillation index with elongation as measured with the Cambridge array for the extragalactic source 3C273 together with a theoretical curve for a double source where each component is 0.35 arcsec in diameter [318]. The Cambridge work also indicates that the spectrum of the irregularities can be represented by a simple Gaussian over the range 5×10^{-3} to 3×10^{-2} km^{-1}, which implies scale lengths of between 30 and 200 km. The solar wind velocities are of the order of 400 km s^{-1}.

Electrons in the interstellar medium also scatter transmitted radio waves and, in theory, it should be possible to use observations of interstellar scintillations (ISS) to establish upper limits as small as 0.001 arcsec for the angular diameters of compact radio sources [326]. To date, the only sources which have been shown to scintillate because of interstellar electron density irregularities are the pulsars (which were discovered in IPS observations!) [360].

Scattering from interstellar electrons broadens the images of compact sources by up to 0.1 arcsec at metre wavelengths. The pulse shapes from pulsars are also broadened by the small time delays introduced by the

Fig. 4.18 Scintillation index as a function of elongation, ε, for the strong QSO source 3C 237. The solid line is a theoretical curve for a symmetrical double source, each component of which has a diameter of 0.35 arcsec. (Published with permission from [318].)

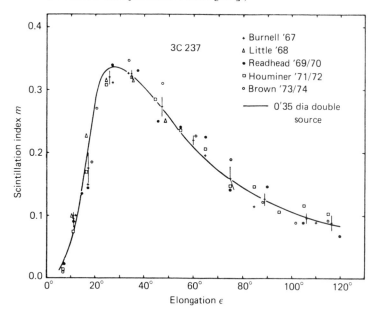

different path lengths between the pulsar and the observer caused by small angle scattering.

Ionospheric birefringence

The refraction of radio waves passing through the ionosphere can be separated into two components: one is caused by the spherical stratification of the electron density, and the other is due to horizontal electron density gradients [89, 317, 337, 423]. The Earth's magnetic field causes birefringence for the two opposite senses of circular polarisation. The expressions for the angles of refraction, R_o and R_e, of the ordinary and extraordinary rays incident at an angle i, perpendicular to the top of the ionosphere for radiation of frequency ω, are

$$R_o = 7.45 \times 10^{-5}((X)/(1+Y))\sigma(\sin i)/(\cos^3 i) \quad \text{rad} \tag{4.16}$$
$$R_e = 7.45 \times 10^{-5}((X)/(1-Y))\sigma(\sin i)/(\cos^3 i) \quad \text{rad} \tag{4.17}$$

where

$$Y = F/\omega \tag{4.18}$$

and

$$X = (f/\omega)^2 \tag{4.19}$$

where F is the electron gyro-frequency corresponding to the component of the Earth's field along the ray path, f is the plasma frequency, and σ is the thickness of the ionosphere in km defined by the ratio $n/(10^3 N)$ (where n is the total number of electrons in a 1 m^2 vertical column and N is the maximum electron density which corresponds to the plasma frequency f).

For a frequency of 22 MHz, $Y \ll 1$, and so the difference $R_e - R_o$ can be approximated by

$$R_e - R_o = 1.49 \times 10^{-4} X Y \sin i/\cos^3 i \quad \text{rad} \tag{4.20}$$

Some typical values for the variables are: $n = 10^{17}$ electrons m^{-2}, $N = 4 \times 10^{11}$ electrons m^{-3}, $F = 1$ MHz. The dependence of both the refractive index and the difference $R_e - R_o$ on ω^{-3} means that birefringence is largely unimportant at centimetre wavelengths but a very significant effect at metre wavelengths. Indeed, at metre wavelengths, the extraordinary and ordinary rays will separate by many tens or hundreds of metres before they reach the ground and they would be of quite different phase. For this reason receivers for reflecting antennae are designed to detect only one sense of circular polarisation.

5

Spectrographs

Only X-ray and gamma-ray detectors have sufficient energy resolution to provide a spectrum of the incoming radiation directly. Even in the X-ray region, the detector energy resolution (see Chapter 7) is insufficient for many astrophysical problems. At longer wavelengths a monochromator is necessary.

Filter photometry is discussed in Chapter 7. This chapter covers some general considerations in the design and choice of spectrographic instruments for use with multi-channel detectors or for operation in a multiplex mode.

Grating spectrographs [34, 351]

Efficiency in the presence of seeing

For optical telescopes in space with diffraction limited optics, the angular width, $\delta\theta_{spec}$, of the entrance slit of a grating spectrograph can be set to the diffraction limit of the grating aperture,

$$\delta\theta_{spec} = \lambda/A \tag{5.1}$$

where A is the illuminated width of the grating. This optimal slit subtends an angle of

$$\delta\theta_{tel} = \lambda/D \tag{5.2}$$

at the focus of the telescope with a primary of diameter D, which is also the diffraction limit of the telescope. The focal length of the spectrograph camera is chosen such that the line spread function of the slit projected onto the detector is adequately sampled.

Unfortunately, for ground based astronomy with large telescopes, the value of $\delta\theta_{tel}$ is much smaller than the diameter of the seeing disc and little radiation would enter the spectrograph from a stellar image if the slit were set to λ/D. For example, $\delta\theta_{tel} = 0.03$ arcsec for a 3.6 m telescope primary at 500 nm, while the seeing disc is typically 1 arcsec.

The diffraction-limited condition is normally relaxed to allow, simultaneously, a large enough entrance slit for the majority of the starlight to pass into the spectrograph while still sampling the line spread function adequately with the detector. The linear dispersion on the detector is controlled by the scale of the spectrograph optics and the choice of angular dispersion.

Figure 5.1 shows, schematically, an all-reflecting spectrograph fed by a telescope primary of diameter D and focal ratio f (focal length/D) [351]. For proper operation the telescope exit pupil should be imaged on the grating by placing the grating at the collimator focus [182]. The spectrograph camera is essentially a Schmidt telescope (see Chapter 3). The grating lies close to the centre of curvature of the spherical camera mirror and either the aspheric corrector is a transmitting element used in a single or double pass mode in front of the collimator, or the aspheric corrections are included in the figure of the collimator. The latter is preferable as they are achromatic.

If the spectrograph camera focal ratio is F, and the sampling interval of the detector is k, then, in the presence of seeing α (radians), the percentage of starlight entering the spectrograph is

$$\text{Efficiency} = (s/fD\alpha) \times 100 \text{ per cent} \tag{5.3}$$

where s is the width of the spectrograph entrance slit and, for convenience, the image is assumed to be square.

This simplified treatment assumes that all of the starlight falling on the primary is contained within α and that the energy is uniformly distributed within the image. Also, it is assumed that the length of the slit exceeds $fD\alpha$. For optimum sampling, the width of the entrance slit as imaged or projected in the spectrum should be at least twice the detector resolution

Fig. 5.1 Optical schematic of an all-reflecting grating spectrograph indicating the various quantities used in the discussion of spectrograph efficiency.

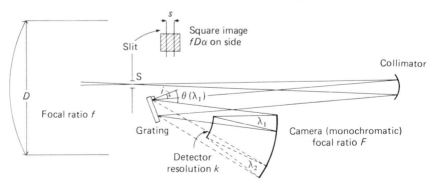

(see section on optimum sampling in Chapter 4). Consequently,

$$s = 2kf/rF \tag{5.4}$$

where [123]

$$r = \cos i/\cos \theta(\lambda) \tag{5.5}$$

i is the angle of incidence between the optical axis of the collimator and the normal to the grating, and $\theta(\lambda)$ is the angle of diffraction at wavelength λ measured from the grating normal, as shown in Figure 5.1 (i and θ are positive when measured on the same side of the grating normal). The term r corrects for the fact that the beam cross-section on the grating is generally elliptical. If the grating is used in an interior order then $r = 0.8$ would be typical, while if used in an exterior order $r = 1.2$ would be a more likely value. Hence, after substituting for s in (5.3),

$$\text{Efficiency} = (2k/rD\alpha F) \times 100 \text{ per cent} \tag{5.6}$$

(the coefficient is 0.4 if k is in micrometres, D in metres, and α in arc seconds).

Since the light gathering ability of the telescope is proportional to $\pi D^2/4$, the speed of the combined telescope/spectrograph system can be defined as

$$\text{Speed} = (2k/rF)M^{0.5} \tag{5.7}$$

where M is the coefficient of surface brightness as defined in equation (4.15).

It is clear from equation (5.6) that the speed of the system increases only linearly with the telescope diameter when a significant fraction of starlight is lost at the spectrograph slit. This emphasises why large telescopes must be put at sites with the best possible seeing if they are to be used effectively for spectroscopy (the case for direct photography has already been made in Chapter 4) [52].

The ideal situation for preserving the efficiency of a telescope/spectrograph combination would be to allow the scale of the spectrograph (including the detector) to increase linearly with the scale of the telescope. This implies that both the ratio k/D and F would remain constant and, in order to retain the same spectral resolution the camera focal length (and hence the scale of the spectrograph) would also vary as k. The principal limitation to this process is the size of available gratings, the largest has a ruled diameter of about 30 cm. At some observatories mosaics of four gratings are used to relieve the size limitation [324].

There has been less effort to increase k with increasing telescope aperture, as this implies a loss of spectral resolution or spectral range.

There is also a tendency for the cosmic-ray background and other noise sources to increase with the area of the individual sensing elements. Spatial resolution varies from detector to detector but adopting, for convenience, a value of $16\,\mu m$ gives, from (5.6), a spectrograph efficiency $= 200/rF$ per cent for a 4 m telescope with one arc second seeing.

The value of r depends on the adopted angular dispersion and the particular grating used. In any event, r can be considered to be approximately 1, which requires that the camera focal ratio must be of order unity if light is not to be lost at the spectrograph slit. This explains why considerable effort goes into the design of fast spectrograph cameras for major ground based telescopes [56].

Image slicers

For stars which are bright compared with the sky background, image slicers have been used very effectively to increase the amount of starlight passing through the spectrograph slit. There are two main types of image slicer [211], one designed by Bowen [313], the other by Richardson [324], and they are shown in Figures 5.2 and 5.3, respectively.

Fig. 5.2 Optical arrangement of a Bowen image slicer. (Published with permission from [313].)

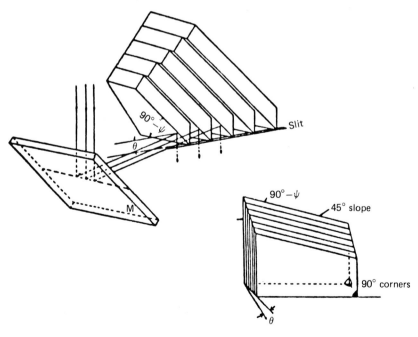

In the Bowen slicer the seeing disc is intercepted by a series of inclined, displaced mirrors, each of which directs a slice of the image onto the slit. The slices cannot be superimposed but are placed end to end, thereby requiring a longer slit than the original image. Although only one slice can be in focus the degree of defocussing of the other slices is not normally important.

Starlight enters the Richardson image slicer through a cylindrical lens which produces orthogonal line images (see Figure 5.3). The horizontal image passes through the entrance slot separating the aperture mirrors while the vertical image coincides with the entrance slit separating the slit mirrors. The field lens immediately behind the slit focusses the entrance slot onto the grating of the spectrograph. Light which does not pass directly through the entrance slit is redirected to other slices on the grating by multiple reflections from the slit and aperture mirrors. The mirrors are separated by their common radius of curvature while the centres of curvature of the aperture mirrors coincide with the edges of the slit mirrors. Light incident on a slit mirror is repeatedly reflected through the centre of curvature at the edge of that particular mirror, which means that a slice of the light passes through the slit on each reflection while the untransmitted light must land on the other slit mirror. The slices are stacked on the grating as shown in Figure 5.4.

The improvement in spectrograph efficiency is given by multiplying equation (5.6) by N, where N is the number of slices imaged on the grating. Similarly, the telescope speed (in equation (5.7)) increases directly as N.

Extended sources

Spectrograph efficiency is independent of both the seeing and the telescope aperture but depends on the inverse of both the telescope and camera focal ratios when the source is extended. *This is only true when angular resolution does not need to be preserved along the slit.* In long slit spectroscopy the telescope exit pupil is re-imaged onto the grating by a lens immediately behind the slit or by some other optical arrangement. If the slit width and length are s and l, respectively, and the variables all have the same definitions as in the previous section then the angular projection of the slit on the sky is

$$\text{Proj.} = sl/(fD)^2 \quad \text{sr} \tag{5.8}$$

The total number of photons falling through the slit per second is given by

$$\text{Flux} = \pi s l n D^2/4(fD)^2 = \pi s l n/4f^2 \tag{5.9}$$

which is independent of D, where n is the source irradiance in photons s^{-1} m^{-2} sr^{-1}. Substituting for s from equation (5.4) gives the flux of photons through the slit as

$$\text{Flux} = \pi l n k / 2 r f F \tag{5.10}$$

i.e. the throughput is proportional to f^{-1} as well as F^{-1}. It is also clear that by using a coarse detector in a large scale instrument the flux of photons through the slit increases linearly as k.

The above arguments do *not* apply to the spectroscopy of galaxies and other sources where angular resolution must be preserved along the slit. In 'imaging' spectroscopy the analysis is the same as for a point source except that, in this case, image slicers cannot be used. This means that good seeing is essential.

Spectral resolution and grating characteristics
The grating equation is

$$m\lambda = d(\sin i + \sin \theta) \tag{5.11}$$

where m is the grating order, d is the spacing of the grating rulings, i is the angle of incidence and θ is the angle of diffraction, both measured from the grating normal. Both i and θ are positive when measured on the same side of the grating normal.

Fig. 5.3 (a) The optical arrangement of a Richardson image slicer, and (b) a cut-away drawing of an actual slicer. (Published with permission from [324].)

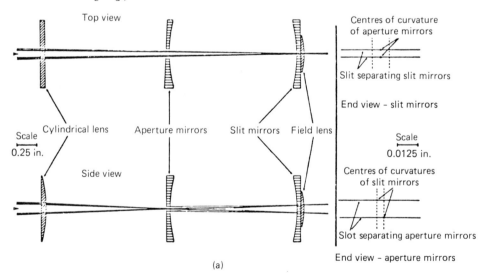

(a)

Figure 5.5 shows the grating efficiency of a commercially available plane grating measured in orthogonal polarisations, one parallel to the rulings, the other normal to them. The individual grooves are tilted at the blaze angle which favours light going into a single order. The blaze angle normally corresponds to a Littrow configuration where $i = \theta =$ the blaze angle. The measurements in Figure 5.5 were made with i and θ differing by 8°. The blaze wavelength changes slowly as this latter angle is increased but this is not an important effect in low orders. Holographic gratings

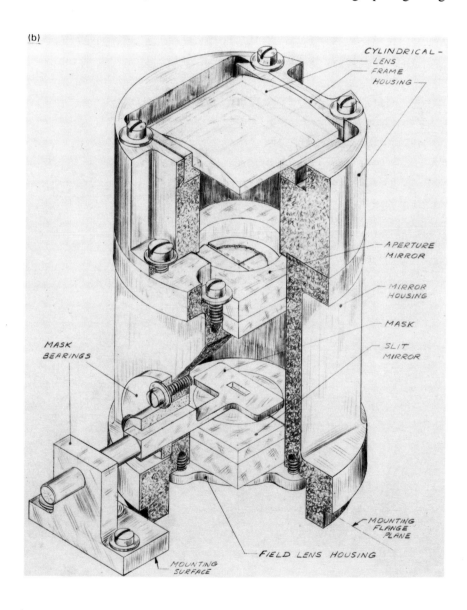

(b)

tend to have a lower overall efficiency than ruled gratings but they have a wider spectral range, less structure in the efficiency curve, and the potential for larger areas than ruled gratings [256].

The linear dispersion (sometimes known as the reciprocal dispersion), $d\lambda/dx$, is

$$d\lambda/dx = (d\cos\theta)/mL \text{ nm mm}^{-1} \qquad (5.12)$$

where $L\ (=FA)$ is the camera focal length.

The spectral resolving power, R, is defined as $R = \lambda/\delta\lambda$, where $\delta\lambda$ is the required spectral resolution (FWHM of the line spread function) such that

$$\delta\lambda/2.5k = (d\lambda/dx) = (d\cos\theta)/mFA \qquad (5.13)$$

where the coefficient 2.5 is introduced for optimum sampling.

Fig. 5.4 Different arrangement for the optical slices on the spectrograph collimator. (Published with permission from [324].)

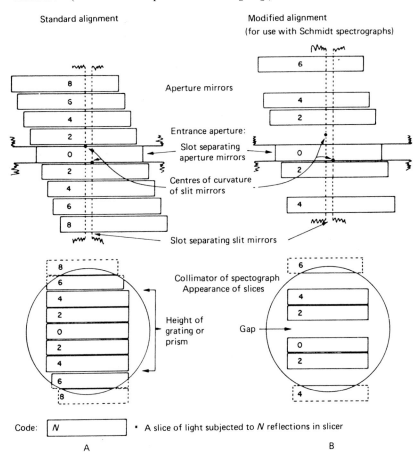

There are physical restraints on each of the variables on the right-hand side of equation (5.13). For a large telescope F must be of order 1 in order to maintain high efficiency as discussed in the previous section. The grating aspect, $d \cos \theta$ must be $> \lambda$, otherwise grating efficiency is reduced. This latter condition limits $1/d$ in the visible spectrum to about 1800 lines per mm for a first-order spectrum. The maximum value of A is 30 cm for single, commercially available ruled gratings, but is doubled in a mosaic of four gratings in some coudé spectrographs [324]. A high resolution, or a more compact spectrograph (smaller value of L) can only be achieved by using orders of $m > 1$.

As a rule of thumb the useful range of a spectral order is given by $\lambda(b)/m$, where $\lambda(b)$ is the blaze wavelength. However, as can be seen in Figure 5.5 the width of the order, $\lambda(b)$, and the grating efficiency are different for the two polarisations. Further, the steep drop in efficiency on the short wavelength side of the curve is normally accompanied by an inflection which corresponds to a Wood's anomaly. The latter are difficult to calibrate for accurate spectrophotometry.

At any given angle of diffraction the grating equation is satisfied by several different values of wavelength and of m. If the radiation from

Fig. 5.5 First order efficiency of a plane grating measured close to the Littrow condition in orthogonal polarisations (parallel and perpendicular to the rulings). Courtesy of Bausch and Lomb.

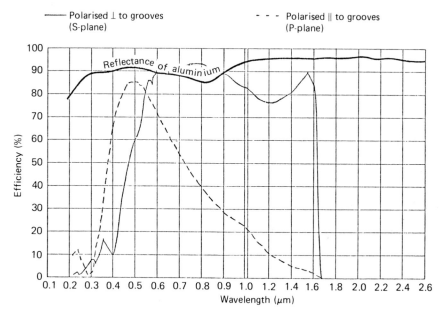

unwanted, overlapping orders falls within the range of sensitivity of the detector they must be filtered out. The filter is usually referred to as an order sorting filter which is normally a blocking filter which blocks all radiation shorter than a certain wavelength. For this reason it is easier to operate in the first order where the only overlap for most detectors is of the easily blocked blue/UV second order into the visible/red region. Second or higher order spectra in the blue/UV are usually seriously contaminated by the visible/red of lower orders. Even in the first order, for spectrographs operating in the vacuum ultraviolet, radiation scattered from longer wavelengths may introduce a strong background for a detector which is not solar-blind (for a further discussion of this problem, see Chapter 7).

Echelle spectrographs [256, 351]

A grating spectrograph operating in a low grating order produces a long, narrow spectrum. Photographic plates can be made of almost any length and, when thin enough and the radius of curvature is greater than about 25 cm, they can be bent to match the focal plane. With the introduction of flat, rectangular multi-channel detectors it is no longer possible to exploit such long spectra and field flatteners must often be introduced to preserve image quality. It is not easy to mount large spectrographs at the Cassegrain focus, consequently it is not possible to use large values of L. The problems of size, rectangular format, and high resolution are largely solved with the use of an échelle spectrograph.

The surface of an échelle grating has a series of steps rather than a series of rulings. A section of a grating and the relevant angles and dimensions are shown in Figure 5.6. If the depth of the steps is d, measured along the step at the grating blaze angle, and the step spacing in the plane of the grating is a, then the grating equation becomes

$$m\lambda = 2d + 2a \sin \theta(b) \tag{5.14}$$

in the Littrow mode. Blaze angles are typically in the 60° to 75° range. The value of the blaze angle is critical for the proper operation of an échelle grating [67]. Normally m is of the order of 50 to 100 and a is coarser than the spacing for plane gratings being of the order of 100.

High resolution is obtained by operating in a high order but at the expense of the limited spectral range in each order and strongly overlapping orders. A single order can be selected with a narrow band filter or, more commonly, the various orders can be separated with a weak cross disperser and imaged together on a single, two-dimensional detector.

One of the most successful uses of an échelle spectrograph has been on the International Ultraviolet Explorer (IUE) satellite [40]. The detectors are described in Chapter 8. The telescope is a 45 cm Ritchey–Chrétien operating at $f/15$. A simplified view of the spectrograph optics is shown in Figure 5.7 together with a perspective view of the short wavelength camera. Figure 5.8 shows the format of the spectra. The échelle grating is cross-dispersed by a concave grating which also acts as the camera mirror. In the low dispersion mode the light to the échelle is intercepted by a plane mirror which gives the single low dispersion spectrum of the cross disperser.

The spectra are in a trapezoidal format which was designed to match the active area of the detectors. The width of the orders in wavelength, $\Delta\lambda$, decreases with increasing values of m. In consequence, the spectra become shorter and closer together. The individual spectra are inclined to the direction of dispersion. Apart from the peak wavelength in each order, light for any given wavelength is divided between two spectral orders. For IUE, the resolving power in the échelle mode is $R = 1.2 \times 10^4$.

There are several limitations of échelle spectrographs in addition to the points made above. There is a continuous curvature in the grating efficiency within any order which must be carefully calibrated. There is a high level of scattered light between orders. Crowding of the higher orders can complicate calibration of the latter and also restricts the slit height for long slit spectroscopy.

Fig. 5.6 An échelle grating in section showing the angles and dimensions used in the text.

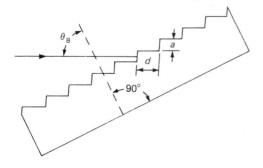

X-ray and infrared

The diffraction grating technique is used successfully over a wide range of wavelengths from the X-ray [116, 153, 224, 350] to the infrared. In the case of the latter the spectrograph optics must be refrigerated to reduce background [401]. Further, multi-channel infrared detectors with sufficiently low noise are not yet generally available. In the infrared wavelength regime (1 to 2.5 μm) in which detector noise exceeds that expected from the combination of the telescope and the sky background, a

Fig. 5.7 Arrangement of the two International Ultraviolet Explorer (IUE) spectrographs and the layout of one of them. (Reproduced from [40].)

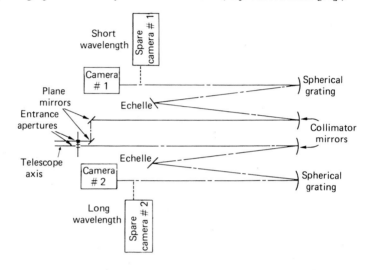

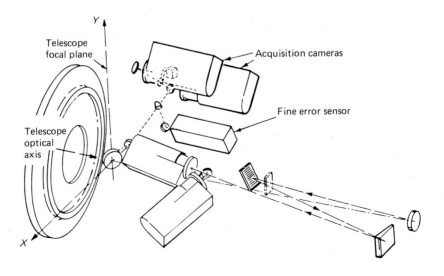

careful choice has to be made between a conventional spectrograph and a multiplexed device such as the Michelson interferometer described below. The question of detector noise, telescope background, and cosmic background is discussed in Chapter 7.

Fig. 5.8 Format of the high dispersion échelle IUE spectra as imaged on the long and short wavelength detectors. (Reproduced from [40].)

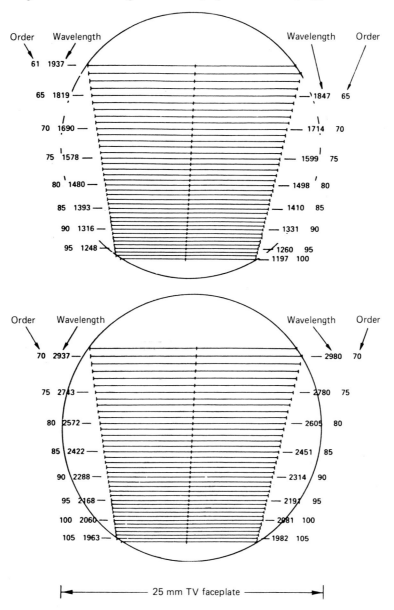

Wide field spectroscopy

In-focus, direct photography with either broad or narrow band filters is equivalent to low resolution spectroscopy for all of the objects on the plate, with the number of spectral resolution elements equal to the number of filters used and the resolution equal to the bandwidths of the filters. Several programmes are now under way using as many as 60 narrow band filters to cover the optical spectrum. The use of filters has the considerable advantage of suppressing unwanted skylight from other spectral regions but suffers from the inefficiency of having to observe with the filters successively rather than simultaneously.

A number of other techniques exist for taking simultaneous spectra of many or all of the objects in the field.

Objective prism, grism, and grens

In common with many other Schmidt telescopes the UK Schmidt telescope (described in Chapter 3) at Siding Spring in Australia has an objective prism [298]. It is 1.26 m in diameter and can be mounted over the aspheric corrector plate. The apex angle is 44 arcmin giving a dispersion of 240 nm mm^{-1} at 430 nm and the prism can be rotated to any orientation. Unwidened spectra of objects as faint as 20 magnitude can be obtained in exposures of about one hour. Normally the spectra are widened by using a tilting block image displacer in front of the autoguider.

Such low dispersion spectra are used for spectral, if not luminosity, classification, for the detection of peculiar objects, or the detection of objects with emission lines such as quasi-stellar objects. Single spectra cannot be used to estimate other than very large radial velocities without including a filter to introduce an absorption feature in the spectrum to act as a wavelength calibration.

Fehrenbach introduced a technique for measuring radial velocities with an objective prism [133]. Pairs of spectra are taken on the same plate with the prism rotated through 180° between exposures. Separations are measured between lines in one spectrum of a pair and the same lines in the other spectrum. This is also done for a number of stars of known radial velocity in the field which act as standards. Differences between the displacements measured for pairs of lines in the programme star spectra and those for a standard star of the same spectral type are attributed to a difference in radial velocity. For this work dispersions of between 8 and 11 nm mm^{-1} are employed with spectra taken under conditions of good seeing (see Chapter 4).

Prisms have the disadvantages of large mass and nonlinear dispersion in the blue and ultraviolet. For larger telescopes where a dispersing element over the entrance pupil would be impractical, transmission gratings have been introduced into the converging beam. A simple grating introduces coma. This is not important at large focal ratios but the primary focal ratios of modern, large telescopes are of the order of 3.

For a grating at a fixed distance from the telescope focus the amount of coma introduced by the grating varies linearly with the inverse of the groove spacing, while the astigmatism varies as the square of the reciprocal of the groove spacing. Thus for a given grating the aberrations will be a minimum if the grating is as far as possible from the focus where one can obtain a good dispersion by using a comparatively coarse grating. By introducing a weak prism which deflects light in the opposite direction to the grating order being used, the point of zero coma can be moved to the centre of the spectrum [53]. Such a device is known as a grism. A grism spectrograph is shown in combination with an aperture plate in Figure 5.9.

A variant of the grism is the grens which combines a transmission grating with a lens [142]. Figure 5.10 shows the cross-section of a triplet, wide-field corrector with a transmission grating on the flat surface of the third lens, which is also slightly tilted to remove residual coma [142]. The focal plane is tilted. The characteristics of the system are given at the top of the figure.

Although transmission gratings can be blazed to give high efficiency in a single order, the light from the zero and other orders contributes to the problems of field crowding and sky background, and the spectra suffer from increasing amounts of coma away from the centre of the favoured order. Figure 5.11 reproduces images from two grens plates taken with the 3.6 m CFH telescope. One is of a crowded field and the other taken at high galactic latitude. For the brighter objects it is possible to see the zero order spectra and the increasing comatic flare in the higher orders.

Aperture plates

Multi-slit wide field spectrographs were used for a long time to observe planetary nebulae and H II regions. Where the majority of the radiation was confined to emission lines the multi-slit had the advantage over the single slit spectrograph of giving emission line profiles from a series of strips of the nebula simultaneously.

With an aperture plate the observer can select the objects of interest to him in the field together with a suitable portion of the sky for comparison, while cutting out the rest of the sky and other objects in the field. Figure 5.9 shows an aperture plate with a grism for use at the prime focus of

Fig. 5.9 An aperture plate used in conjunction with a grism on the Kitt Peak National Observatory 4 m telescope to produce low dispersion spectra of preselected objects at the Ritchey–Chrétien $f/7.6$ focus. (Published with permission from [159].)

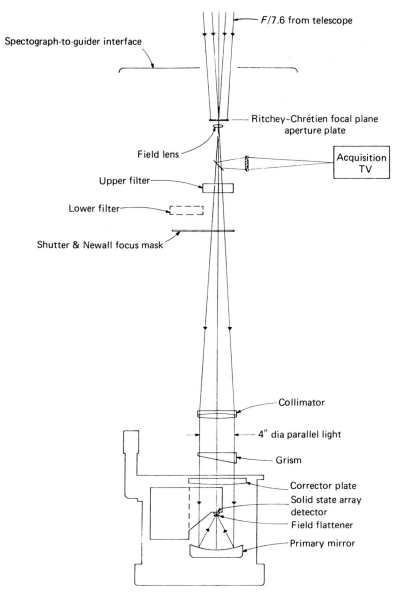

the Kitt Peak National Observatory 4 m telescope [159]. Based on the coordinates of the objects of interest, holes corresponding to 2.5 arcsec diameter are drilled at the appropriate positions in a plate which is inserted as shown in the Figure. The detector is inside the cryogenic camera. The positions of the holes and the rotation angle of the grism are optimised to minimise overlap among the spectra.

Alternate aperture plate designs exist which allow the use of conventional, long slit spectrographs fed by optical fibres from the various apertures [162, 201, 263]. (Optical fibres are discussed further in Chapter

Fig. 5.10 Cross-section of the grens optics used in conjunction with the wide field corrector of the Canada–France–Hawaii telescope (see Figure 3.27). The transmission grating is on the plane surface of the third lens which is slightly tilted to remove residual coma. The focal plane is also slightly tilted. (Published with permission from [142].)

	Green	Blue
Dispersion =	200 nm/mm	100 nm/mm
Blaze =	500 nm (350/850 nm)	430 nm (350 /600 nm)
A wedge =	$0.68° \pm 0.05°$	$1.11° \pm 0.05°$
α tilt =	$0.22° \pm 0.02°$	$0.39° \pm 0.02°$

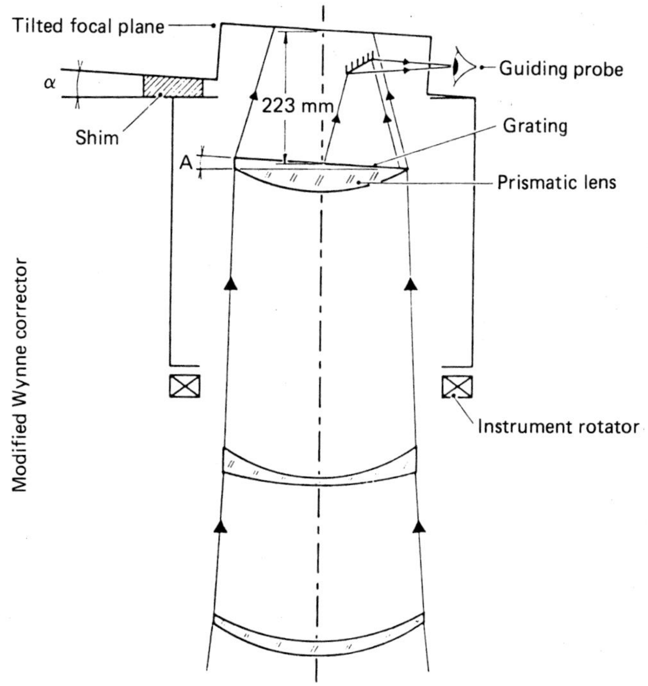

Fig. 5.11 Images from two plates taken with the grism arrangement in Figure 5.10. One plate is from a crowded field and the other taken at high galactic latitude. Photographs courtesy of D. Crampton.

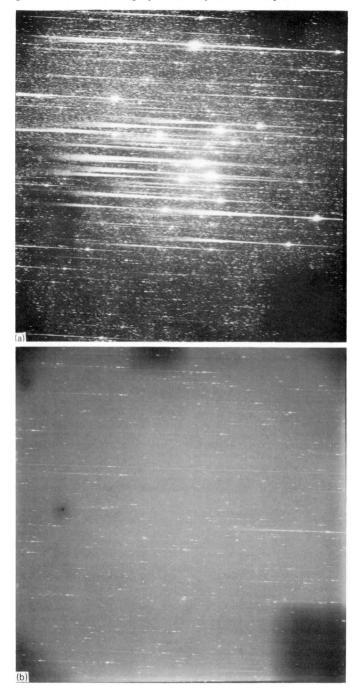

3.) Figure 5.12 is a photograph of the European Southern Observatory 'optopus' for use on the 3.6 m telescope [263]. Up to 54 optical fibres with core diameters of 133 μm guide the light from apertures in the focal plane 'starplate' to the spectrograph where they are formed into a slit. The free ends of the fibres are fed by microlenses which convert the telescope $f/8$ beam to $f/3$ in order to reduce beam dispersion within the fibres. The slit and the lens collimator form a single unit which replaces the normal collimator. Two coherent fibre bundles each of 1 mm diameter (individual fibres 50 μm diameter), with reticles at their inputs, project images of two guide stars onto the intensified CCD for guiding and accurate orientation of the 'starplate'.

Fabry–Perot interferometers [10, 284, 335]

Figure 5.13 is an optical schematic of a Fabry–Perot interferometer. When collimated light (source at infinity) is transmitted by the two partially reflecting parallel plates interference fringes are imposed

Fig. 5.12 The European Southern Observatory (ESO) 'optopus' arrangement for multi-object spectroscopy with a conventional spectrograph. Light from objects isolated by holes in the focal plane 'starplate' is guided to the spectrograph by optical fibers where they are arranged in line as a slit. Guide star images are fed to an intensified CCD by two coherent fiber optic bundles with input reticles. European Southern Observatory photograph, courtesy G. Lund.

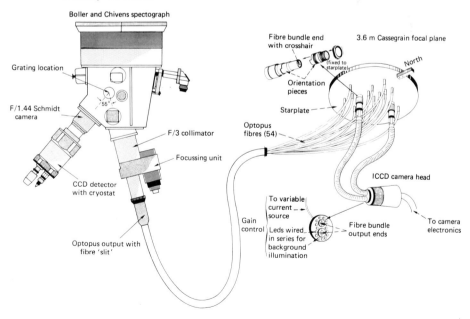

on the image. For a monochromatic, uniform source such as the sky, the interference pattern consists of concentric circles centred on the optic axis. The variation in transmitted intensity with the angle, θ, of the incident radiation, I_{max} is given by

$$I(\theta) = I_{max}(1 + 4r^2 \sin^2(\delta/2)(1 - r^2)^{-2})^{-1} \qquad (5.15)$$

where r is the surface fractional amplitude reflection from a single surface which means that r^2 is the fractional reflection of intensity. For rays incident at angle θ, a plate separation (etalon spacing) t, and refractive index n of the medium separating the plates, the phase difference between successively reflected rays, δ, is given by

$$\delta = 2\pi \cdot 2nt \cos \theta / \lambda \qquad (5.16)$$

such that for maximum transmission

$$\delta_{max} = m \cdot 2\pi \qquad (5.17)$$

where m is an integer, i.e.

$$m\lambda = 2nt \cos \theta \qquad (5.18)$$

Each concentric ring in the interference pattern corresponds to a single value of m.

There is also an angular dispersion associated with the Fabry–Perot because the wavelength of maximum transmission decreases with increasing θ,

$$\lambda = \lambda_0 \cos \theta \fallingdotseq \lambda_0(1 - \theta^2/2) \qquad (5.19)$$

where λ and λ_0 are the wavelengths of maximum transmission at the angles of incidence θ and 0 degrees.

For a uniform source such as an extended nebula emitting monochromatic light from an emission line, the ring pattern is convolved with the intensity distribution in the source. If, further, there is a range of

Fig. 5.13 Optical schematic of a Fabry–Perot interferometer indicating the angles used in the text.

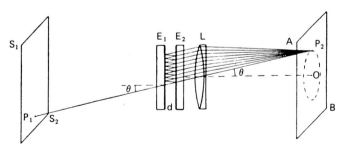

radial velocities in the source, corresponding segments of the ring pattern will be displaced according to equation (5.18). By varying either n or t it is possible to scan the full image of the nebula in frequency and thereby map it both in radial velocity and in intensity. Normally the emission line of interest is isolated with a narrow band interference filter (see Chapter 7) or a second etalon as described below.

The intensity maxima sharpen as r^2 increases as shown in Figure 5.14. The finesse, F, is defined as the ratio of the full width at half of maximum intensity of an order $\delta\lambda$ to the interorder separation, $\Delta\lambda$,

$$F = \Delta\lambda/\delta\lambda = \pi r/(1 - r^2) \tag{5.20}$$

F depends only on the reflectivity r. The closer r is to unity the sharper the transmission peaks.

The theoretical resolving power, R, is given by

$$R = \lambda/\delta\lambda = mF \tag{5.21}$$

For normal incidence, $m = 2nt/\lambda$, corresponding to the separation of the plates, t, in numbers of wavelengths. Consequently, the resolving power of the system depends directly upon the plate separation. As an example, consider the resolving power of a system with $\lambda = 500$ nm, $r = 0.9$, $n = 1$, and $t = 20$ mm. From the value of r it follows that $F = 14.9$, giving a value for R of 1.2×10^6.

With the introduction of low attenuation, all-dielectric, multi-layer reflection coatings (see Chapter 7 for a discussion in connection with interference filters), the Fabry–Perot technique has become widely used. Early observations were confined to bright objects because metallic coatings introduce a strong attenuation.

The technique is used in two principal areas. Both involve a controlled variation of the optical path length between the plates either by using

Fig. 5.14 The variation of angular transmission of a Fabry–Perot with r^2, the percentage reflected intensity from the coatings of the parallel plates.

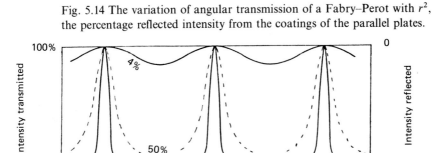

piezoelectric spacers, or by varying the density and hence the refractive index of the gas between the plates [49, 320, 321, 343]. Wide field velocity and line intensity surveys of galaxies and nebulae carried out with large telescopes use a focal reducer to achieve collimation and to reduce the field scale to match the area of the detector.

Some imaging Fabry–Perot systems are used as tunable monochromators for restricted fields of view. If the field of view is restricted to $\delta\theta$, where

$$\delta\theta = (2\delta\lambda/\lambda)^{0.5} \tag{5.22}$$

then this covers the full-width half-maximum transmission of the central or 'bull's-eye' fringe. Associated with this is the change in wavelength across the field due to the Fabry–Perot angular dispersion (see equation (5.19)) [334, 418].

The great advantage of the Fabry–Perot is the possibility of observing line intensities over the whole image simultaneously. This advantage relative to the use of a long slit, conventional spectrograph is offset by the need to obtain many exposures with a range of spacings to examine a single line. For a system with high finesse the bandpass is narrow and little radiation reaches the detector. For modern scanning Fabry–Perot systems, a photon counting detector of high spatial stability is essential and it must be coupled to a data acquisition system which can accumulate data from as many as forty individual images corresponding to the different etalon spacings.

Two or more Fabry–Perot etalons having different interorder separations are used in the 'Pepsios' system [203, 265]. This arrangement in combination with a wide band filter can isolate a single etalon order without the use of a narrow, isolating filter.

Although the principle of the Fabry–Perot interferometer is valid at all wavelengths there are no appropriate dielectric coatings available for $\lambda < 300$ nm. A recent aircraft borne experiment in the 30 to 200 μm region used 40 lines mm^{-1} mesh of electroformed nickel stretched across two optically flat steel rings to act as the reflecting plates [6]. To scan, one screen was moved relative to the other with a motor driven cam.

Michelson interferometer [102, 173, 267, 284, 327, 349]

The Michelson interferometer is the most frequently used of the more general class of Fourier transform spectrometers. A schematic for the optics of a simple scanning Michelson interferometer is shown in

Figure 5.15. Light from the telescope divides at the beam splitter. Half continues to the moving mirror (B) and the other half is directed to the fixed mirror (A). After reflection the beams again divide at the beam splitter and recombine at the detectors, one of which is on the opposite side of the beam splitter from A, the other in the direction of the incoming beam. The beams from A and B interfere because of the difference in path length in the two optical trains. If the path difference is x then the phase difference, $\phi(k)$, for radiation of wave number k is

$$\phi(k) = 2\pi k x \qquad (5.23)$$

If the amplitude of the incoming radiation is A the amplitudes in beams A and B will each be $A/2^{0.5}$ for a perfect beam splitter. After the second interaction with the beam splitter an amplitude of $A/2$ from each of beams A and B is directed to each detector. The amplitude of the radiation received at the single detector (D) on the opposite side of the beam splitter from A is

$$A(\text{D}) = A/2 + (A/2)\exp(2\pi i k x) \qquad (5.24)$$

The intensity of the radiation at the detector, $I(\text{D})$, is then

$$I(\text{D}) = (I/2)(1 + \cos(2\pi k x)) \qquad (5.25)$$

Fig. 5.15 Optical schematic of a simple, scanning Michelson interferometer.

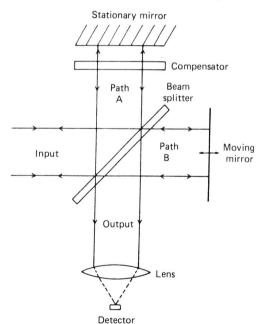

For dielectric beam splitters the radiation intensity at the detector in the direction of the input, $I(I)$, compliments that at D such that

$$I(I)=(I/2)(1-\cos(2\pi kx)) \qquad (5.26)$$

With metallic beam splitters $I(D)$ and $I(I)$ are in phase. With the latter there is a significant absorption loss. There is little absorption loss with dielectric beam splitters but they introduce phase distortion through dispersion.

From equation (5.25) it can be seen that the output signal has a constant term and one which varies sinusoidally with x. The rate of variation of the sinusoidal term depends directly on the velocity of mirror B and the wave number, k. Clearly, if the incoming radiation is not monochromatic the modulation frequency will be different for different wavelengths. In general, infrared detectors respond only to the a.c. component of the incident signal which also makes them insensitive to light scattering within the instrument. The a.c. component is given by $f(x)$, where

$$f(x)=\frac{1}{2}\int_{\lambda_2^{-1}}^{\lambda_1^{-1}} I(k)\exp(2\pi ikx)dk \qquad (5.27)$$

where λ_1 and λ_2 are the bandpass limits normally imposed by prefiltering of the incoming radiation.

For a properly adjusted interferometer, radiation of all wave numbers is in phase at $x=0$ where the intensity is I. As the path difference increases, the signal drops to a mean level of $I/2$.

In practice, a series of interferograms is accumulated with the cycle time for the acquisition of each interferogram being less than the expected rate of change in the background or source intensity caused by atmospheric variations and instrumental instabilities. Normally both outputs are recorded and their combined signals act as a convenient 'volume control' to monitor variations in the integrated signal level.

It is necessary to take an inverse Fourier transform of the interferogram in order to display the spectra in the frequency domain,

$$f(x)=\int f'(k)\exp(2\pi ikx)dk \qquad (5.28)$$

and

$$f'(k)=\int f(x)\exp(-2\pi ikx)dx \qquad (5.29)$$

where $'$ denotes a Fourier transform.

If the total excursion of mirror B is a distance a, then the total range of x is $2a$. The potential resolution of the interferometer, $R=\lambda/\delta\lambda$, can be derived from the general expression [284]

$$R = nN \tag{5.30}$$

where n is the order of interference, and N is the number of interfering beams. In the case of the Michelson interferometer $N = 2$, and $n = 2a/\lambda$, hence

$$R = 4a/\lambda \tag{5.31}$$

The loss of resolution caused by non-normal incidence is discussed in the next section.

The finite range of the scan and the finite sampling interval Δx of the interferogram introduce a periodic, sync function of half-width λ/R response in the wavelength domain. The sync function is analogous to the instrumental profile of the conventional slit spectrograph. It can be apodised to a single lobe (at the expense of wavelength resolution) by weighting the interferogram in a manner similar to that described in Chapter 3 for the tapering of radio antennae [284, 349].

To avoid aliasing or periodicity in the sync function, the spectral range of the incoming radiation, $\Delta\lambda$, must be restricted by filters, such that

$$\Delta\lambda \leqslant \lambda_1^2/2\Delta x \tag{5.32}$$

where Δx is the sampling interval, and λ_1 is the minimum wavelength transmitted.

Alternatively, if the maximum and minimum wave numbers of the incident radiation are k_1 and k_2, respectively, then the interferogram should be sampled at intervals of at least $\frac{1}{2}(k_1 - k_2)$ in x.

Cube corner or cat's-eye reflectors rather than plane mirrors are normally used in modern interferometers and considerable attention must be paid to calibration of mirror position and overall alignment, particularly for instruments with large values of a. Mirror position is usually determined by counting fringes from an associated laser interferometer. A full discussion of interferometer design and data analysis is beyond the scope of this book. Readers interested in details should consult one of the primary references [173, 267, 284].

The Fellgett and multiplex advantages

It was pointed out in Chapter 2 that shot noise in the detected radiation and detector readout noise are the principal noise sources in incoherent detection. Instruments operate most efficiently when photon shot noise exceeds detector readout noise. This is discussed quantitatively in Chapter 7. For infrared detectors, readout noise expressed in equivalent photons is much higher than for optical detectors.

With the Michelson interferometer, all of the incident radiation from the full bandpass of the instrument reaches the detector rather than radiation from a single spectral element. This is known as multiplexing. For a given wavelength resolution, $\delta\lambda$, and bandpass, $\Delta\lambda$, the degree of multiplexing, N, is given by

$$N = \Delta\lambda/\delta\lambda \tag{5.33}$$

to give an advantage of $1/N$ in exposure time when compared with a single channel spectrometer.

For a sufficiently broad bandpass, the photon shot noise will exceed the detector readout noise. As a result the detector operates at a higher detective quantum efficiency (DQE is discussed in Chapter 7) than an equivalent, multi-channel spectrograph of the same resolution. This is known as the Fellgett advantage. With continual improvements in detector performance, the equivalent bandpass in which the photon shot noise exceeds the readout noise has become smaller for a given source strength.

There is an important additional source of noise from the sky and telescope background radiation included in the projection of the detector on the sky. It increases inversely as the irradiance of the source and increases rapidly for wavelengths $> 2\,\mu m$. Once the background photon shot noise exceeds the readout noise in equivalent photons from the detector, the Fellgett advantage is lost. This point is discussed in Chapter 7.

With the development of multi-channel infrared detectors (discussed in Chapter 8) it is possible to exploit the additional advantage of the Michelson interferometer over a conventional spectrometer by taking spatially resolved interferograms over a two-dimensional field [333]. An interferogram can be generated for each detector element. The wavelength resolution of such an instrument can be varied in the subsequent data analysis by binning samples to give anything between λ/R to a direct photograph using the prefilter bandpass, $\Delta\lambda$.

The Jaquinot advantage

The range of incident angles introduced by the finite field of view of an interferometer degrades its resolution compared with one used entirely at normal incidence, particularly at short wavelengths. Jacquinot pointed out that Ω, the acceptance solid angle, was restricted, such that [216]

$$\Omega \leqslant 2\pi/R \qquad (5.34)$$

For a circular field of θ radians radius, this gives

$$\theta = (2/R)^{0.5} \qquad (5.35)$$

He used this result to demonstrate that Fourier spectrometers can accept radiation from a larger solid angle than, say, a long slit spectrograph for the same spectral resolution.

In modern instruments Ω is significantly increased by a variety of techniques over the value given in (5.34) while retaining the equivalent resolution [102, 284].

Autocorrelation spectroscopy at radio frequencies [2, 19, 57, 99, 104]

Early radio receivers in astronomy used frequency scanning and multi-channel filtering techniques [308]. Modern radio receivers tend to be of the autocorrelation type. They are a radio analogue of the optical

Fig. 5.16 The delay line of an r.f. autocorrelation spectrometer. Signals are fed in from either end and sensed at the tap points. See text. (Published with permission from [19].)

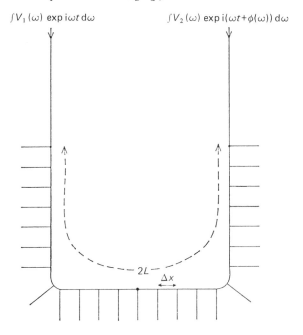

$\int V_1(\omega)\, \exp i\omega t\, d\omega$ $\int V_2(\omega)\, \exp i(\omega t + \phi(\omega))\, d\omega$

$2L$

Δx

Fourier transform spectrometers discussed in the previous section. The autocorrelation receiver enjoys many of the advantages of the Michelson interferometer in terms of flexibility, of resolution, and of large band width. Most are operated in a digital mode [2, 104].

The principle of an autocorrelation system is illustrated in Figure 5.16. Signals from two separate telescopes are fed into either end of a delay line of length $2L$. Detectors are attached at equal distances Δx along the line. At the mth detector from the midpoint the time delay between the signals is $2m\Delta x/c$. If the inputs at the two ends of the line are

$$\text{Inp}_1 = \int V_1(\omega)\exp(i\omega t)d\omega \tag{5.36}$$

and

$$\text{Inp}_2 = \int V_2(\omega)\exp(i(\omega t + \phi(\omega)))d\omega \tag{5.37}$$

where $\phi(\omega)$ is a phase difference introduced by a switch in one arm, then, ignoring attenuation in the line, the output from the detector at m is

$$\text{Out}(m) = \int V_1 V_2(\cos\phi \cdot \cos(2m\omega\Delta x/c)$$
$$+ \sin\phi \cdot \sin(2m\omega\Delta x/c))d\omega \tag{5.38}$$

The two terms in the brackets correspond to the in-phase (cosine) and quadrature (sine) outputs of a conventional phase-switched interferometer (see Chapter 3 for details) multiplied by the sinusoidal phase-switching frequency. The set of $2m-1$ outputs from the detectors is proportional to the cross-correlation function of the input signals sampled at equal intervals of the time delay, $2\Delta x/c$. The Fourier transform of the set gives the power spectrum of the correlated signal over a band width of $c/4\Delta x$ with a resolution in frequency of $c/4(m-1)\Delta x$ Hz. An analogous arrangement can be used with the correlation interferometer discussed in Chapter 6.

6

Dilute apertures

It was pointed out in Chapter 2 that the angular resolution of simple telescopes can be improved by the use of an extended baseline formed by two or more telescopes. The improvement in resolution is achieved either by triangulation (gamma rays) or, interferometrically, by measuring the degree of coherence of the source.

The interplanetary gamma-ray burst network [92, 93, 125]

Gamma-ray bursts of nonsolar origin were first detected by four of the Vela group of Earth-orbiting satellites [230]. Subsequently, an interplanetary network was established of nine spacecraft having gamma-ray detectors on board which has detected some ten gamma-ray bursts per year. The bursts generally show a very rapid rise time of the order of milliseconds followed by pulsations with periods of seconds and an overall decline in intensity over several hundred seconds. The sudden onset of the bursts followed by an identifiable pulsation signature allows accurate timing at each spacecraft from which it is possible to determine the direction of the source by triangulation. The principal errors come from uncertainties in timing and knowledge of the spacecraft positions and motions since, for accurate triangulation, the position and timing of a spacecraft must be correlated with Universal Time (UT).

Table 6.1 lists the spacecraft, their positions, and the various onset times for an event detected on 5 March, 1979 which at that time was the most intense burst of nonsolar high energy photons ever recorded [92, 93, 125]. The interplanetary separations of the spacecraft provided baselines of up to 500 light-seconds.

The coordinates of the spacecraft were known in geocentric coordinates. Positions in stellar or nebular catalogues are quoted in a heliocentric system and so corrections were necessary for (a) the velocity of the Earth relative to the Sun (the aberration of the gamma-ray source) and (b) precession of the equinoxes. Clock errors due to relativistic effects caused by relative velocity or differing gravitational fields amounted to no more than 0.5 ms [35, 93].

The difference between the onset times (in UT) for the burst at each pair of spacecraft defined a small circle or band on the sky normal to the baseline formed by the two spacecraft and with a width which was proportional to the combined timing errors and inversely proportional to their separation. The empirically derived timing precision was between 30 and 60 ms although the design accuracies were in the 1 to 5 ms range. Figure 6.1 shows the positional error box of about 1×2 arcmin generated using the error bands from the three baselines formed by the PVO, ISEE 3, and Helios B observations.

The supernova remnant N49 in the Large Magellanic Cloud (LMC) falls within the error box and, since the most likely source for such a burst is a neutron star, the coincidence appears to confirm the validity of the triangulation technique. (If indeed the source is at the distance of the LMC (55 kpc) and the burst of gamma rays was emitted isotropically then the radiated energy of the burst was between 10^{36} and 10^{37} J and, at its peak intensity, the source was emitting between 10^{37} and 10^{38} W which is a factor of 10^{11} to 10^{12} more than the Sun!)

Two-beam interferometry
Intensity interferometry [175, 176, 177, 178]

The correlation of wave noise (see Chapter 2) in the signals received at two telescopes from a coherent source was first demonstrated at radio frequencies (where photon noise is unimportant) using baselines of up to 12 km [218]. However, the greatest effort to exploit the property

Table 6.1 *Onset time and ephemeris information for nine spacecraft*

Spacecraft	R (km)	RA (1950)	Decl. (1950)	T_0 (s)
Helios B	69 959 568	47.177 5	17.643 5	57 115.126
ISEE 3	1 649 622	333.443 6	−6.823 6	57 124.903
PVO	152 355 112	303.927 366	−18.890 375	57 118.027
Vela 5B	127 631	352.80	−49.85	57 124.536
Vela 5A	119 403	270.11	−4.44	57 124.962
Vela 6A	117 617	82.83	−6.92	57 124.639
Venera 11	141 344 485	288.710 74	−24.662 49	57 099.150
Venera 12	142 993 285	287.024 14	−24.269 67	57 104.354
Prognoz 7	71 471	310.521	35.796	57 125.002

of correlated wave noise from partially coherent sources has been directed towards measuring stellar diameters at optical wavelengths.

The observational arrangement is shown schematically in Figure 6.2 [175]. Two telescopes of collecting area A are separated by a distance d. Radiation is fed to the photomultipliers (see Chapter 7) D_1 and D_2 and their output currents pass through low pass filters which have a bandpass, Δf, of between 1 and 100 MHz before being multiplied in C.

From the point of view of technique it is perhaps easier to consider the wave picture rather than the tendency of photons to bunch within the coherence time as outlined in Chapter 2. Each point of a source radiates over a range of frequencies and quite independently of any other point. By Fourier analysis the radiation from any point can be represented as a superposition of sinusoidal components. If $E'\sin(\omega't+\phi')$ and

Fig. 6.1 The positional error box of about 1×2 arcmin estimated for the gamma-ray burst source detected on 1979 March 5 in the direction of the Large Magellanic Cloud. The box is formed by the intersection of the small circles defined by the differential onset times of the burst at pairs of the satellites listed in Table 6.1. Two pairs from PVO, ISEE 3, and Helios B give the smaller errors with Venera and Prognoz being the third pair. (Published with persmission from [125].)

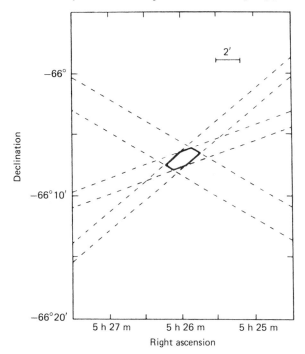

Fig. 6.2 (a) The component layout of the intensity interferometer. (b) The principle of an intensity interferometer. (Published with permission from [175].)

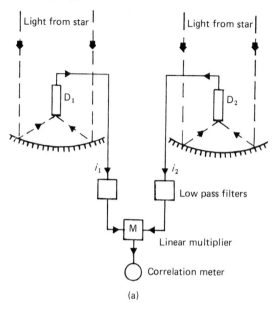

(a)

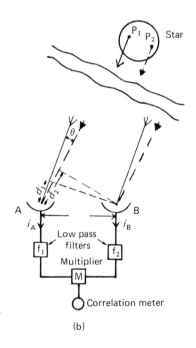

(b)

$E'' \sin(\omega''t + \phi'')$ are the Fourier components of the electric field received at D_1 from two points 1 and 2 on the source then the output current from D_1 due to these components is proportional to the intensity or the square of the net electric field,

$$i_1 = \varepsilon(E' \sin(\omega't + \phi') + E'' \sin(\omega''t + \phi''))^2 \tag{6.1}$$

where ε is a constant for the telescope and photomultiplier.

The same two Fourier components also illuminate D_2 and generate a current,

$$i_2 = \varepsilon(E' \sin(\omega't + d'/c + \phi') + E'' \sin(\omega''t + d''/c + \phi''))^2 \tag{6.2}$$

Expanding (6.1) and (6.2) gives

$$
\begin{aligned}
i_1 = \tfrac{1}{2}\varepsilon(&(E'^2 + E''^2) - (E'^2 \cos 2(\omega't + \phi') + E''^2 \cos 2(\omega''t + \phi'')) \\
&- 2E'E'' \cos((\omega' + \omega'')t + (\phi' + \phi'')) \\
&+ 2E'E'' \cos((\omega' - \omega'')t + (\phi' - \phi'')))
\end{aligned} \tag{6.3}
$$

$$
\begin{aligned}
i_2 = \tfrac{1}{2}\varepsilon(&(E'^2 + E''^2) - (E'^2 \cos 2(\omega'(t + d'/c) + \phi') \\
&+ E''^2 \cos 2(\omega''(t + d''/c) + \phi'')) \\
&- 2E'E'' \cos((\omega' + \omega'')t + \omega'd'/c + \omega''d''/c + (\phi' + \phi'')) \\
&+ 2E'E'' \cos((\omega' - \omega'')t + \omega'd'/c - \omega''d''/c + (\phi' - \phi'')))
\end{aligned} \tag{6.4}
$$

The first terms in each expansion are the normal d.c. output which is proportional to the total flux falling onto the photomultipliers. In the intensity interferometer these are rejected by the low pass filters. The second and third terms correspond to the second harmonics ($2\omega'$, $2\omega''$), and the sum frequencies ($\omega' + \omega''$), both of which lie outside the filter bandpass.

Both of the fourth terms correspond to the beat frequency ($\omega' - \omega''$) which does lie within the filter bandpass and, in theory, they are the only ones multiplied at **M**. Although these two components have the same frequency they differ by ($\omega'd'/c - \omega''d''/c$) in phase and their product is

$$c(d) = \varepsilon^2 E'^2 E''^2 \cos((\omega/c)(d' - d'')) \tag{6.5}$$

after making the approximation $\omega' = \omega'' = \omega$. From the geometry of Figure 6.2, equation (6.5) can be rewritten as

$$c(d) = \varepsilon^2 E'^2 E''^2 \cos(2\pi\theta d/\lambda) \tag{6.6}$$

where θ is the angular separation of the two points on the source.

When equation (6.6) is integrated over all possible pairs of points on the source, all possible pairs of Fourier components within the optical bandpass, and over all difference frequencies which lie within the bandpass, one obtains the following simple result [175],

$$c(d)/c(0) = V(d)^2 \tag{6.7}$$

where $V(d)$ is the fringe visibility as defined in equation (2.4), and $c(0)$ is a constant for the equipment. The relative forms of $V(d)$ and $V(d)^2$ are shown in Figure 2.2 for a uniform circular source.

The degree of wave-noise correlation depends on the difference in phase between the low frequency beats at the two detectors. Consequently, the coherence length associated with the beat frequency is, by equation (2.9)

$$\Delta l = c/\Delta f \tag{6.8}$$

where Δf is the highest frequency passed by the filters (10^8 Hz) which is one million times less than that of the light vibrations (10^{14} Hz). This means that the limit on the difference in path length from the source to each of the two telescopes is only of the order of 1 cm (i.e. a few thousand wavelengths) unlike the classical Michelson interferometer where the stability must be a few micrometres. This property makes the technique largely immune to the (nondispersive) effects of atmospheric turbulence.

The correlated noise component involves only fourth order terms in equation (6.5) and, as pointed out in Chapter 2 from equation (2.13), this implies, for bright stars in the optical region, effects of the order of a few parts in 10^7 of the total noise. The consequent need for a large photon flux and the relaxed tolerance on path length led Hanbury Brown and Twiss to adopt a 'light-bucket' concept for their two-telescope intensity interferometer at Narrabri. Each telescope consisted of 252 hexagonal 38 cm diameter mirrors for a total collecting area of 30 m^2 with a focal length of 10.85 m. The reflectors, on trucks, were moved around a circular track of 180 m diameter in order to follow a star in azimuth. They were tilted about the horizontal to follow it in elevation. The amplified outputs from the two photomultipliers were fed by cables to the central control building.

The bandpass of the optical filter was 12 nm centred at 443 nm. Figure 6.3 shows the correlation of $c(d)/c(0)$ with baseline for Sirius A obtained in a total of 203 hours observing. The angular diameters of 32 single and multiple stars were measured before the installation was finally dismantled. Although the instrument was insensitive to seeing effects it was necessary to observe for very long periods in order to accumulate a sufficient signal-to-noise in the correlation. Figures 6.4 and 6.5 summarise the limitations.

Figure 6.4 shows the minimum aperture diameter required to achieve a signal-to-noise of 3 in one hour as a function of stellar magnitude. At a given apparent magnitude, cooler stars must subtend a larger angle, θ, than stars of higher temperature (see discussion in Chapter 2).

Fig. 6.3 Correlation as a function of baseline for Sirius A. The points are observations and the full line is the prediction for a model atmosphere which includes limb darkening and an angular diameter of 5.9×10^{-3} arcsec. Results on the three longest baselines are also shown on an expanded scale. (Published with permission from [175].)

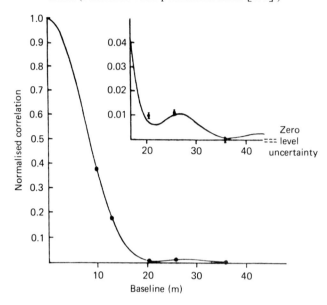

Fig. 6.4 The minimum reflector diameter in metres necessary for an intensity interferometer to resolve a star of a given magnitude. (Published with permission from [175].)

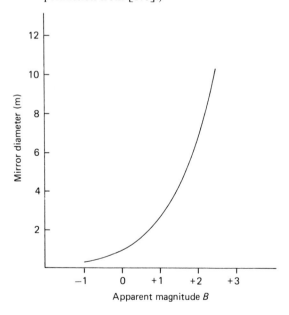

Consequently, as the telescope aperture is increased, light from the cooler stars loses coherence more rapidly than for the hotter stars or, put another way, the mirror size derived from Figure 6.4 exceeds the maximum baseline at which the radiation is still coherent. The effect is summarised in Figure 6.5 and implies that the technique cannot be used for stars with surface temperatures of less than about 5000 K.

The Michelson stellar interferometer
Optical

The elements of coherence theory based on the classical dual-beam interferometer are developed in Chapter 2 where it was shown that the resolving power of a full aperture can be doubled along one axis by only using two apertures at the ends of a diameter. Michelson exploited this concept by using aperture separations of up to 6 m on the 2.54 m Mount Wilson telescope with the optical arrangement shown in Figure 6.6 [288].

When the beams from the two mirrors, M_1 and M_2, are adjusted to give superimposed images at the Cassegrain focus, O, fringes will be formed for sufficiently monochromatic light from a coherent or partially coherent

Fig. 6.5 The maximum achievable signal-to-noise ratio for a two-mirror intensity interferometer in one hour as a function of stellar temperature. (Published with permission from [175].)

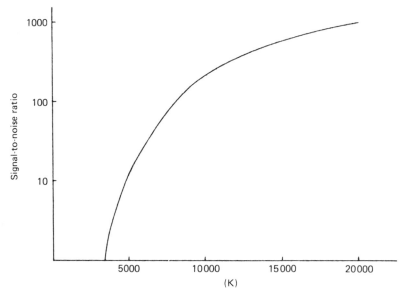

source provided that the path lengths are equal within a coherence length (equation (2.9)). The mirror separation can be varied until the fringes first disappear, which, for a star, will be for the condition $d = 1.22\lambda/\theta$, where θ is the angle subtended by the stellar disc. The equality of the two path lengths is maintained to within a few micrometres (the coherence length) by use of the compensator in one beam.

Detail in the fringe pattern suffers the same seeing-induced phase distortions as images formed by the full telescope. Since the mirrors are similar in size to r_0, the turbulence coherence length, the predominant problem is a rapid relative motion of the images from each aperture. Figure 6.7 shows several 20 ms exposures of the fringes formed by Vega in a two-beam interferometer [240]. The fringes were formed using a 40 nm bandpass at a common coudé focus of two 0.25 m telescopes on a 12 m baseline and detected by a sensitive television camera.

The relative wandering of the two images is obvious but the fringes are always clearly present. The telescopes were altitude-altitude mounted on a north-south line [241]. The coudé focal ratios were 3000. This design, which uses the ground substrate for support rather than a cantilever or beam arrangement, is highly stable against flexure and is the prototype for a much larger multi-telescope array being constructed near Grasse, France [164, 241, 242].

It is obvious that in any long exposure the integrated effect of image motion would wash out the fringe pattern. By using photon-counting area

Fig. 6.6 Principle of the Michelson stellar interferometer. Beams from the two mirrors M_1 and M_2 are adjusted to give superimposed images at the Cassegrain focus O. The path lengths are equalised by adjusting the optical delay in C_2.

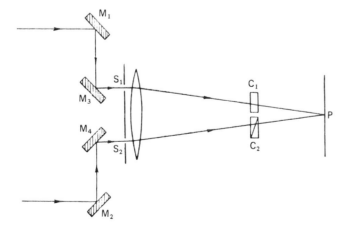

Fig. 6.7 A series of two-beam interferometric fringes of Vega obtained with a 400 nm bandpass at the common $f/3000$ Coudé focus of two 25 cm telescopes. The telescopes were altitude-altitude mounted on a 12 m north-south baseline. (Published with permission from [240].)

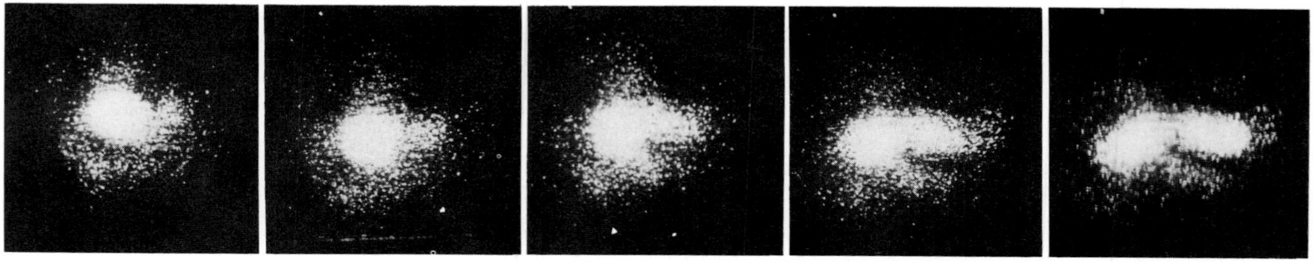

detectors which register the position of detected photons within each frame, it is possible to calculate a time-averaged autocorrelation function of the fringe pattern from the autocorrelation of the photon distribution in each frame. This is similar to the analysis used in speckle interferometry.

The use of Michelson interferometers to measure stellar diameters has, to date, been largely demonstrative of the technique rather than used for any extended programmes. The breech-piece interferometer, on the other hand, was used extensively for objective measurements of double star separations and position angles.

Figure 6.8 shows the optical arrangement of the interferometer built and used by Finsen [137]. By having the pupil close to the eyepiece the separation and orientation of the slits (which form the apertures of the interferometer) can be carefully, and rapidly, changed by hand while continuously observing the fringe pattern. This is not possible with apertures at the objective.

The width of the slits can be varied depending on the brightness of the stars. For double stars of modest separation the interferometer is rotated to give maximum fringe visibility such that the line of centres of the slits is perpendicular to the stars. The slit separation is varied to give minimum fringe visibility in the orthogonal direction which allows a separation to be estimated (provided the double is close enough to give a minimum visibility at some slit separation). If the stars are equally bright the fringe visibility becomes zero at the appropriate slit separation.

Infrared incoherent detection

Figure 6.9 is a schematic of a portable interferometer which can be mounted at the Cassegrain focus of different telescopes [279]. It uses a single InSb detector and a filter bandpass of 0.7 μm centred at 5 μm. The mask (1) isolates two circular portions of the converging telescope beam.

Fig. 6.8 Optical arrangement of a breech piece interferometer used to measure double stars. The width, separation, and orientation of the slits can be conveniently changed by hand to give the maximum fringe visibility. (Reproduced from [137].)

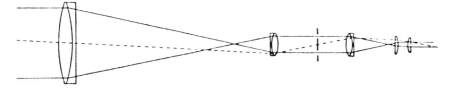

The beams are intercepted by the plane mirrors (2) and directed by the roof prism (3) to the dichroic beam-splitter (4), where the optical radiation is transmitted to an image monitor (10, 11), while the infrared radiation is deflected into the Dewar (5) where, after transmission through the filter (6), fringes are formed on the mask (7). The mask has three slots to match the pitch of the interference fringe pattern shown schematically in (9). The detector is shown at (8).

The mirrors (2) are mounted on a stack of piezoelectric crystals and the fringe pattern is phase-switched at 40 Hz by moving one of the mirrors sufficiently to introduce a $\lambda/2$ path difference. This changes the phase of the fringe pattern by π (the technique is analogous to phase switching for radio frequency interferometry described in Chapter 3). Figure 6.10 shows the results of letting the fringe pattern from Arcturus drift across the mask. The record shows the phase-sensitive output from the detector $I(0) - I(\lambda/2)$. Such a drift scan makes it possible to measure the fringe visibility, $V(d)$ (see equation (2.4)), directly.

Fig. 6.9 Schematic of a portable infrared interferometer mounted at a telescope Cassegrain focus. The components are described in the text. (Published with permission from [279].)

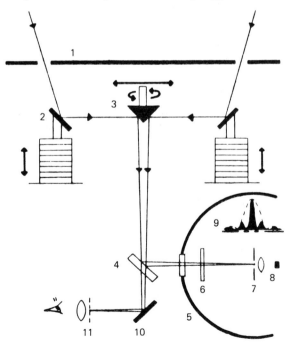

A range of baselines is achieved by using telescopes of different apertures. When the optical seeing exceeds 4 arcsec the phase distortion is too rapid (>40 Hz) at 5 μm for the formation of stable fringes. The fringes in Figure 6.10 were obtained when the visual seeing was 1 arcsec with aperture diameters of 0.2 m and a separation of 1.02 m projected at the telescope mirror. The radiation was sufficiently monochromatic ($\Delta\lambda = 0.7\ \mu$m, $\lambda_2 = 5\ \mu$m) that, according to equation (2.6), there should have been approximately 11 fringes which, in fact, are observed. Even with good seeing the apparent fringe visibility must be corrected for loss of contrast caused by atmospheric effects, instrument and telescope flexure, and setting and tracking errors. Together these introduce some 5 to 20 per cent uncertainty in the adopted value of $V(d)$.

Infrared coherent detection

In coherent detection the two beams are detected independently and the corresponding electronic signals are either added or multiplied together to generate fringes. The bandpass in heterodyne detection is inherently much smaller than that used in incoherent detection which means that the coherence length is orders of magnitude greater. The relaxation in coherence is made at the expense of a much weaker signal.

The components and the operation of a heterodyne interferometer operating at 10 μm are shown in Figure 6.11 [221]. The heterodyne receiver is described in more detail in Chapter 7 [310]. The interferometer uses the solar auxiliary telescopes of the McMath Solar Telescope at KPNO. Each telescope has a flat heliostat of some 0.4 m^2 area separated

Fig. 6.10 An interferogram of Arcturus obtained with the interferometer in Figure 6.9 during a slow scan in right ascension. (Published with permission from [279].)

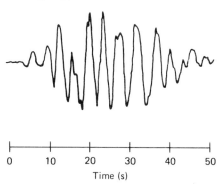

0   10   20   30   40   50

Time (s)

by 5.5 m on an east-west baseline. The flats feed off-axis paraboloidal mirrors which focus the stellar images onto the receivers. At the focus of each telescope a high-speed Ge:Cu photoconductor mixes the source radiation with the beam from a 1 W CO_2 laser. The lasers are phase-locked with a 5 MHz difference in frequency to avoid interaction between them.

The separate radio frequency signals from each photomixer preserve the original phase and amplitude of the incoming radiation. Each photomixer output is amplified and passed through a stepped coaxial cable delay line to compensate for changes in path length as the source moves across the sky. The 1200 MHz band width of the signals means that the coherence length is several centimetres (in contrast with the bandpass of 10^{12} Hz and coherence length of about 35 μm for the 5 μm incoherent system described above). This gain in path length tolerance is at the expense of sensitivity since the received signal is inversely proportional to the coherence length.

The two appropriately delayed signals are multiplied in the broad band correlator. After removing the 5 MHz offset signal the frequency of the correlator output should be proportional to the rate with which the projected baseline of the interferometer changes as the source moves across the sky. The fringe frequency is $(\phi D \cos d \cos h)/\lambda$, where ϕ is the

Fig. 6.11 Block diagram of the CO_2 heterodyne interferometer described in the text. (Published with permission from [221].)

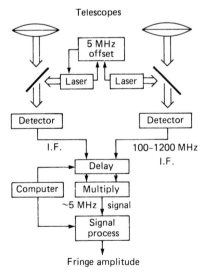

Earth's rate of rotation, d is the declination of the source, and h is the source hour angle (which increases linearly with time). The amplitude of the output is directly proportional to the fringe visibility.

Radio interferometers

An important advantage of coherent detection is the preservation of the relative phases of the incoming signals. In modern radio interferometric systems the signals from the different telescopes are multiplied together, or correlated, to give normalised fringe amplitudes which vary as $\cos \phi$ rather than the $(1 + \cos \phi)$ fringes which result from addition as described in Chapter 3.

Figure 6.12 shows, schematically, two radio telescopes connected to form a correlation interferometer. A delay can be introduced in one arm to compensate for the difference in the wavefront arrival times $(\Delta t = (d \sin \theta)/c)$ at each telescope. The associated phase delay, ϕ, for a source at an angle θ and wavelength λ is

$$\phi = 2\pi(d/\lambda) \sin \theta \quad \text{rad} \tag{6.9}$$

The inputs from the two telescopes are multiplied together in two ways

Fig. 6.12 Elements of a two-beam correlation interferometer. The associated cosine and sine fringe patterns for a point source are also shown.

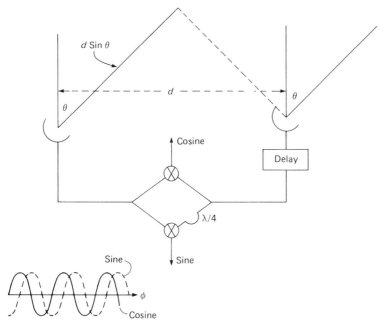

to give a 'cosine' and a 'sine' output. The sine multiplication is carried out after one input has been delayed, in phase, by $\pi/2$ in an extra $\lambda/4$ length of cable.

The output voltages from the two telescopes contain correlated contributions V_1 and V_2 corresponding to the degree of coherence in the source. In addition there are uncorrelated components from extended incoherent structure in the source and background plus contributions from the noise generated in each system. The time average of the voltage products from the two telescopes will be equal to the product of their correlated components only, provided the product is averaged over a sufficiently long period. It can be shown that the 'cosine' output is given by [86]

$$\overline{V_1 \cdot V_2} = \sqrt{\overline{V_1^2} \cdot \overline{V_2^2}} \cos \phi \qquad (6.10)$$

where overbars indicate time average, and the 'sine' output is given by

$$\overline{V_1 \cdot V_2} = \sqrt{\overline{V_1^2} \cdot \overline{V_2^2}} \sin \phi \qquad (6.11)$$

The cosine and sine fringe patterns complement each other as shown in Figure 6.12. If the delay is adjusted to maintain $\phi = 0$ for the direction θ_0 then the normalised cosine output would be unity and the sine output zero for a coherent source in the direction θ_0. If, instead, the coherent source is in the direction $(\theta_0 + \Delta\theta)$ such that $\phi = \pi/2$ then the cosine output would be zero, the sine output would be unity, and $\Delta\theta = \lambda/(4d \cos \theta_0)$. This value of $\Delta\theta$ represents the half-width of either the cosine or the sine fringes.

When the telescopes track the same source the cosine or sine fringes of the interferometer are equivalent to small circles on the sky centred on the interferometer baseline as shown in Figure 6.17. The reception pattern is limited by the angular size of the Airy disc of a single telescope ($1.22\lambda/a$, where a is the diameter of one of the telescopes). The Earth's rotation will carry the fringe pattern across the source unless the time delay is continuously adjusted to maintain $\phi = 0$.

Figure 6.13 shows the geometry of the interferometer [378]. The baseline intersects the celestial sphere at B which has a declination δ and a local hour angle h. The source is at S with declination Δ and hour angle H. The projection of the baseline on the intersection of the plane SOB and a plane tangent to the celestial sphere at S is $d \cos \theta$. This can be resolved into the components, u and v, expressed in wavelength units in the declination and hour angle directions, respectively, where

$$u = (d/\lambda) \cos \delta \cdot \sin(H - h) \qquad (6.12a)$$
$$v = (d/\lambda)(\sin \delta \cdot \cos \Delta - \cos \delta \cdot \sin \Delta \cdot \cos(h - H)) \qquad (6.12b)$$

These equations define ellipses in the *u–v* plane as the baseline rotates with the Earth. For simplicity one telescope can be considered fixed while the other revolves around it with a period of a sidereal day. As seen from the source, the projection of the orbit will, in general, be elliptical.

For a source near the celestial pole the locus in the *uv* plane is a circle of radius d/λ, on the equator the locus is a straight line of length $2d/\lambda$, and for intermediate declinations the eccentricity is cos δ. Unless the source is circumpolar (i.e. δ exceeds the complement of the geographical latitude of the interferometer), it cannot be observed continuously around the ellipse.

Long baseline interferometry [69]

The first radio interferometers had short enough baselines that the separate receiver outputs could be directly connected to a central correlator. Maintaining direct connections over distances greater than a few kilometres is difficult however, and, even at such spacings, many sources are still unresolved.

With the introduction of highly stable atomic frequency standards it is no longer necessary to correlate receiver outputs in real-time. Instead, they can be recorded separately, together with the signal from a local frequency standard, and replayed and correlated later when the two recordings are brought together. The technique is known as long baseline interferometry (**LBI**) and the only restriction on the telescope separation is that the telescopes must be capable of observing sources simultaneously [21, 65, 66, 74, 96, 231, 297].

Data can either be recorded directly as an analogue signal or be converted to a digital data stream before recording [69]. Although the

Fig. 6.13 The spherical geometry of an interferometer showing the angles described in the text. (Adapted from [378].)

techniques are different, the playback principle is the same. Figure 6.14 is a block diagram of a typical analogue receiving system. The rubidium frequency standard provides a 5 MHz signal (accurate to one part in 10^{13}) to control the phase of the first local oscillator and the single side band mixer. The data is recorded on standard high density video tape and the frequency standard is used to precisely phase the TV sync pulse and to provide a time code directly on the audio channel.

The playback scheme is shown in Figure 6.15. Tapes from the two telescopes are mounted on separate tape recorders tied to the same 1 MHz oscillator. The sync pulses are used to coarsely synchronise the recorders and a fine correction, ε, is generated from a comparison of the time codes on the audio channels. An appropriate delay is introduced in one channel before the calibrated signals (freed from the sync pulses) are multiplied in the correlator. The output voltage from the correlator is then a measure of fringe visibility and phase.

The fringe visibility for the Quasar 3C273 is shown in Figure 6.16 for

Fig. 6.14 Components of a recording analogue receiving system suitable for long baseline interferometry. Courtesy of J. Galt.

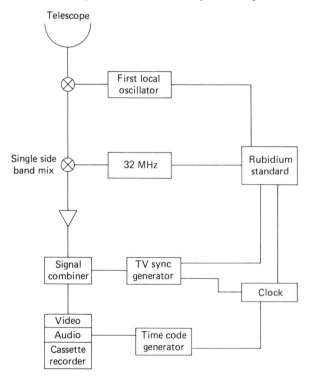

LBI baselines between 0 and $3.5 \times 10^7 \lambda$ [69]. The source is partially resolved at a baseline corresponding to 0.02 arcsec but remains constant at the same level of coherence for all longer baselines. This suggests that the source has two components of roughly equal intensity, one is a halo of 0.02 arcsec diameter while the other is unresolved and less than 0.005 arcsec diameter.

Fig. 6.15 Block diagram of a long baseline interferometer tape playback scheme. Courtesy of J. Galt.

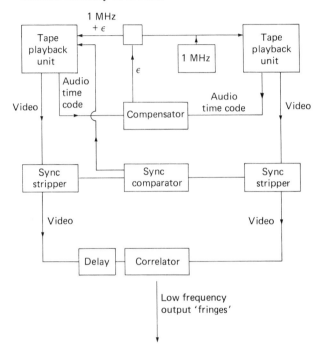

Fig. 6.16 Fringe visibility as a function of baseline for the quasar 3C 273. The source is interpreted as having a resolved halo of 0.02 arcsec and an unresolved core. (Reproduced from [69].)

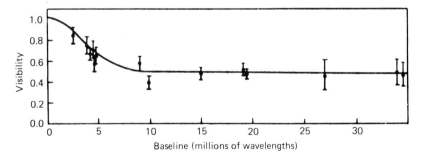

Fringe visibility over a wide range of baselines can be established by using existing radio telescopes at different wavelengths and different projected baselines. A worldwide very long baseline interferometer (VLBI) network has been established [16]. The analysis of such multiple baseline data is discussed in the next section.

Aperture synthesis [12, 62]

The fixed baseline, two-beam interferometer is one of a more general class of multiple and variable baseline interferometers used at radio frequencies. If the delay, $(d \sin \theta)/c$ (see Figure 6.12), is continuously adjusted to maintain $\phi = 0$ in the direction θ_0 then the effective reception pattern of the interferometer consists of two overlapping sets of fringes, one cosine and the other sine, displaced by $\Delta\theta = \lambda/(4d \cos \theta_0)$ with a length of $2.44\lambda/a$ radians where a is the diameter of the individual telescope apertures. Although the centres of these two fringe patterns can be maintained in the direction θ_0 by adjustment of the delay and correct tracking of the telescopes, they rotate with the Earth about the direction θ_0 and have projected baselines in declination and hour angle in accordance with the uv equations (6.12a and b) as shown in Figure 6.17.

It was pointed out above that the time-averaged interferometer output is a function of only the correlated signals from the component telescopes. The cosine and sine outputs are, consequently, proportional to the degree of coherence across each fringe, i.e. source structure of the order of $\Delta\theta$ $(= \lambda/(4d \cos \theta_0))$ or smaller. If the source is observed for six, or preferably twelve, hours, source structure can be mapped in both right ascension and declination over the range of angular frequencies defined by u^{-1} and v^{-1}. When the baseline is altered by moving one telescope, or, by using a pair of telescopes with a different baseline, then the source can be mapped for a different set of u and v values and hence for a different range of angular frequencies.

Where the telescope baselines provide a sufficiently complete coverage in u and in v it is possible to map all of the structure in the source within the limits set by the longest baseline and the diameter of the individual telescopes. If D is the maximum baseline available and d is the smallest relative incremental displacement of the telescopes, then complete sampling of the uv plane requires of the order of D/d spacings in increments of d. The radius of a single telescope aperture, $a/2$, is the smallest useful value for d.

With a single set of either cosine or sine fringes, there would be an ambiguity of 180° in establishing the position of off-axis structure. The simultaneous observation of both the cosine and sine outputs resolves the ambiguity. This point is illustrated below in connection with the observation of the radio source BG 2107 + 49.

An alternative, and more general, description of the aperture synthesis technique can be made in terms of Fourier transforms. The fringe visibility and the phase can be derived from the cosine and sine outputs and, for any given orientation of the interferometer, determine one point on the Fourier transform of the angular intensity distribution in the source. Therefore, by observing with a range of baselines which adequately covers the *uv* plane, it is possible to establish the two-dimensional Fourier transform and to produce a map by inversion [341]. Apart from a lower sensitivity, the only differences between this technique and the use of a

Fig. 6.17 Projection of a two-beam interferometer fringe pattern on the sky.

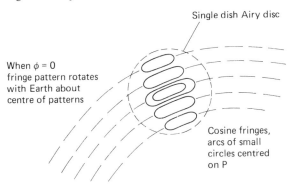

Single dish Airy disc

When $\phi = 0$
fringe pattern rotates
with Earth about
centre of patterns

Cosine fringes,
arcs of small
circles centred
on P

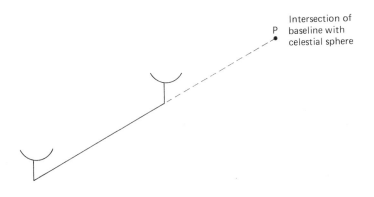

Intersection of
P   baseline with
celestial sphere

very large, filled, aperture are the finite number of points observed and the fact that the points cannot be observed simultaneously.

Interferometers specifically developed for aperture synthesis (sometimes called supersynthesis or Earth rotation synthesis) have telescopes distributed in such a way that permutations of all the possible baselines formed by pairs of telescopes provide a uniform *uv* plane coverage with a minimum of baseline redundancy, or they use a combination of fixed and movable telescopes to provide the necessary range of baselines [20].

Fig. 6.18 The Dominion Radio Astrophysical Observatory supersynthesis array at Penticton, British Columbia. Photograph courtesy of the Herzberg Institute of Astrophysics.

Figure 6.18 is a photograph of a supersynthesis array of the latter type at the Dominion Radio Astrophysical Observatory, Penticton, British Columbia, consisting of four 8.6 m paraboloids [336]. Two of the paraboloids are fixed and two can be moved on a 300 m track, and the interferometer is used for observations of the 21 cm line of neutral hydrogen. The four telescopes are arranged on an east-west line as shown in Figure 6.19 and have a minimum spacing increment, d, of 4.3 m. With these dimensions it is possible to map a 2° area of the sky with an angular resolution of 2 arcmin. The delay is introduced by a cable network in steps of 4 cm.

Fig. 6.19 The arrangement of the four telescopes on the east-west line of the DRAO supersynthesis array. (Published with permission from [336].)

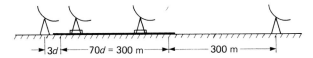

Fig. 6.20 Sets of 21 cm cosine (real) and sine (imaginary) records made over 12 hours, of the galactic source BG 2107 + 49 as observed on three different baselines with the DRAO supersynthesis array in Figure 6.18. Observations in this figure and Figures 6.21 and 6.22 courtesy of P. van der Werf, M. Goss, L. A. Higgs of Dominion Radio Astrophysical Observatory, Penticton.

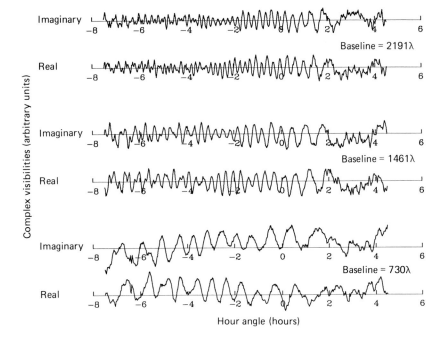

Figure 6.20 shows three sets of cosine and sine records generated with three different baselines over twelve hours of observations at 21 cm of the galactic nebula BG 2107 + 49. The complex visibility shows the relative signal received and zero is indicated by the horizontal lines in each record. The cosine record corresponds to that marked 'real' and the sine record, 90° out of phase, corresponds to 'imaginary'. A final map of the source based on the records from thirty-six different baselines spaced at grating intervals of 81λ is shown in Figure 6.21. There is a resolved source close to the centre of the map and a bright point source to the south.

If there had been only a single, unresolved source at the centre of the

Fig. 6.21 A 'clean' 21 cm map of the source BG 2107 + 49 produced by a 'CLEAN' deconvolution algorithm from the 'dirty' map in Figure 6.22. The synthesised beam (fwhm) is 1(EW) × 1.3(NS) arc minutes and the contour levels are 4, 10, 18, 33, 58, 104, 183, 323, 569 milli-jansky per beam width. Courtesy of the Herzberg Institute of Astrophysics, Dominion Radio Astrophysical Observatory.

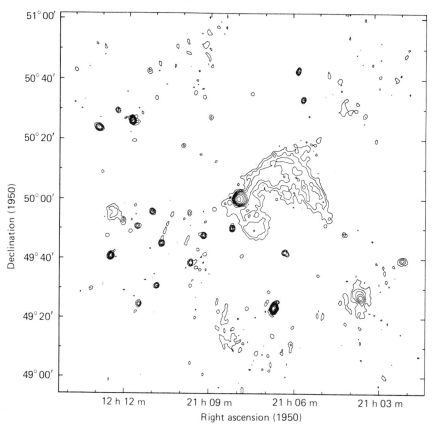

map then the cosine record would have been constant with unit visibility and the sine record zero. For BG 2107 + 49 the record oscillates as the southern point source moves through the sine and cosine fringes with the longer baselines having closer fringe spacings. The central source moves through fewer fringes and produces the lower frequency modulation.

Figure 6.22 shows the initial map formed by inversion from the thirty-six baselines. The under sampling in the uv-plane introduces an obvious 'grating' response. Such 'dirty' maps can be 'cleaned' by interpolation schemes in the Fourier domain [91, 95, 353]. The map in Figure 6.21 was produced by a version of the 'CLEAN' deconvolution algorithm. The synthesised beam (FWHM) is 1(EW) × 1.3(NS) arc minutes.

Fig. 6.22 A 'dirty' 21 cm map of BG 2107 + 49 produced by inversion from complex visibilities at thirty-six baselines (three sets are shown in Figure 6.20) with incremental spacings of 81λ to which the 'grating' response corresponds. Courtesy of the Herzberg Institute of Astrophysics, Dominion Radio Astrophysical Observatory.

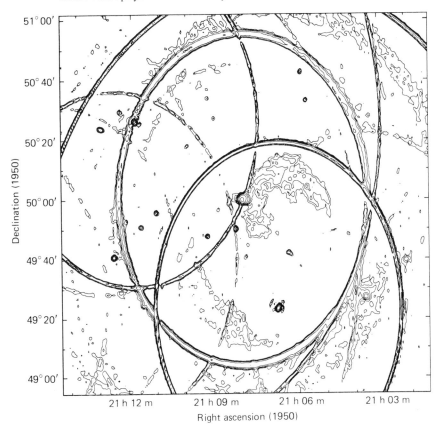

Oscillator drifts, uncertainties in source position and, particularly for long baseline interferometers, phase errors introduced by propagation through the ionosphere can introduce significant phase errors into the cosine and sine outputs. For this reason it is important that a source be observed over 12 hours allowing the baseline to reverse. This largely eliminates systematic phase errors [219, 319].

The very large array (VLA)

The very large array (VLA), near Socorro, New Mexico, is a synthesis instrument which has twenty-seven 25 m Cassegrain telescopes arranged along three 21 km long arms of a 'Y' with nine telescopes in each

Fig. 6.23 The large scale configuration of the Very Large Array stations. (Reproduced from [381].)

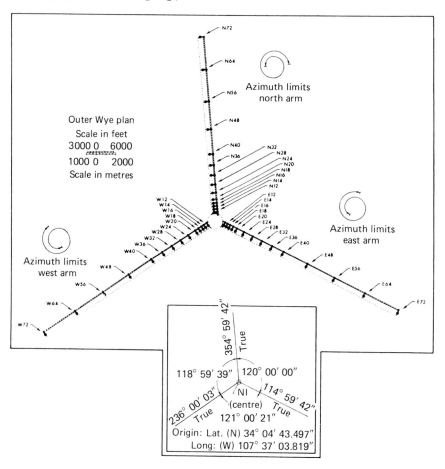

[85, 381]. The operating wavelengths are 1.3, 2, 6, and 18 to 21 cm and there are four standard configurations, A (35 km maximum baseline), B (11 km), C (3.5 km), and D (1 km). The large scale plan of the array in Figure 6.23 shows the outer telescope stations together with the orientation of the arms. Figure 6.24 shows the inner part of the array.

For any given array configuration the distances of the telescopes from the array centre are proportional to $m^{1.716}$, where m is the antenna station number. With this choice of power the mth station on any configuration coincides with the $2m$th station of the next smaller configuration. In this way all four configurations can be handled by 24 stations in each arm.

Figure 6.25 shows the possible uv plane coverage for a series of declinations using all 27 telescopes. For $\delta > 64°$ 24-hour tracking is assumed, while for smaller declinations the tracking is limited by the visibility of the source (elevation limitation). Both the standard (u, v) and the conjugate $(-u, -v)$ ellipses are plotted for a total of 702 full ellipses or elliptical segments.

Sufficient distortion is introduced into the symmetry of the array by inclining the N arm of the Y 5° west of the true north-south line that

Fig. 6.24 The centre of the VLA. (Reproduced from [381].)

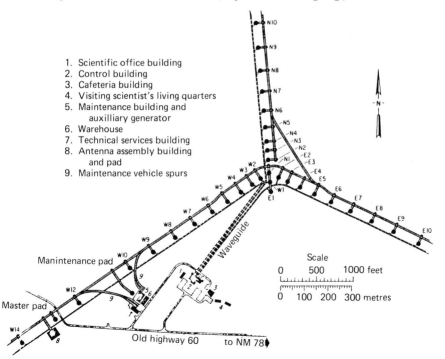

1. Scientific office building
2. Control building
3. Cafeteria building
4. Visiting scientist's living quarters
5. Maintenance building and
    auxilliary generator
6. Warehouse
7. Technical services building
8. Antenna assembly building
    and pad
9. Maintenance vehicle spurs

ellipses from different pairs of telescopes do not overlap. It is possible to map a source adequately from a single 8-hour observation. At 2 cm wavelength the angular resolution in the A configuration is 0.2 arcsec with a field of 2.9 arcmin. Both the resolution and the field scale roughly as the wavelength.

The large number of baselines available at any given instant has meant that the VLA has been particularly effective in providing 'quick looks' at previously unmapped sources. 'Snapshots' taken with observing times of only about five minutes are often sufficiently detailed to allow the observer to make a judgement about the value of continuing to observe a particular source.

Fig. 6.25 *uv* plane coverage for various declinations using all of the VLA's 27 telescopes. (Reproduced from [381].)

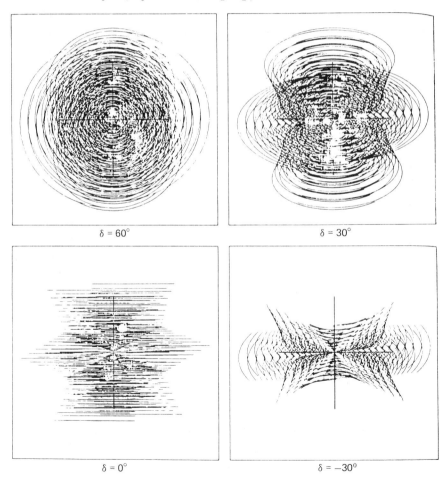

$\delta = 60°$ $\delta = 30°$

$\delta = 0°$ $\delta = -30°$

Occultations

The edge of the shadow cast by an occulting screen which is illuminated by a distant point source will show a classical Fresnel diffraction pattern in monochromatic light. When the source is extended and/or the radiation is not monochromatic the modulation of the interference pattern will be reduced. The measurement of the light distribution at the edge of the shadow can therefore provide some information about the angular distribution of intensity in the source in the direction normal to the occulting edge. If the position of the screen is well known then the edge of the shadow can also be used to define the coordinate of the source in the direction normal to the edge.

Lunar occultations

Before the development of aperture synthesis techniques and speckle interferometry, lunar occultations offered the best opportunity of measuring accurate source positions and angular diameters, particularly at radio wavelengths [235, 300]. The position and the motion of the Moon are known to within a few metres and a few metres per second and the time of occultations can be predicted within a millisecond.

The Moon scans about 10 per cent of the sky during the 18.6-year lunar precession cycle (between declinations $\pm 28.5°$) and scans about one per cent of the sky each year. Generally, a source will be occulted in several successive months and then not again for several years. The geometry changes between occultations and between observing sites. As a result it is possible to carry out some mapping of the source by observing the same occultation from different sites.

The Moon, at a mean distance of 3.84×10^5 km from the Earth, moves across the sky at approximately 0.5 arcsec per second of time. Consequently the lunar shadow travels across the ground at about 900 m s^{-1}. One metre at the surface of the Earth subtends an angle of 5.4×10^{-4} arcsec at the Moon. Although the intensity pattern can be described in terms of the classical Fresnel integrals, the linear scale, s, of the 'shadow bands' or fringes, for radiation of wavelength λ metres, is roughly equivalent to that of the half-wave zones at a distance d behind a straight occulting screen and is given, approximately, by

$$s = (n\lambda d/2)^{0.5} \quad \text{m} \tag{6.13}$$

where n is the fringe number, counting from the shadow edge. A single low order fringe therefore subtends an angle of approximately $7.5\lambda^{0.5}$ arc seconds at the Moon where λ is in metres.

Figure 6.26 shows the theoretical brightness distributions normal to the edge of the shadow for a variety of source diameters at a wavelength of 520 nm. The scale on the ground is indicated in metres. If the radius to the source makes an angle ϕ with the vector of lunar motion, as shown in Figure 6.27, the velocity of the fringe pattern is $900\cos\phi\,\mathrm{m\,s^{-1}}$.

From equation (6.13) it is clear that the scale of the shadow bands is proportional to $\lambda^{0.5}$. The approximate dwell time, t, of a fringe on the telescope is

$$t = (\lambda d/2)^{0.5}/v\cos\phi = (15/\cos\phi)\lambda^{0.5}\quad\mathrm{s}\qquad(6.14)$$

where λ is in metres. In order to resolve the fringes they should be sampled spatially every $s/10$ metres, or every $(1.5/\cos\phi)\lambda^{0.5}$ seconds. This implies a sampling frequency of 67 kHz at 0.1 nm (X-rays), 1 kHz in the optical region, and 0.7 Hz at 1 metre.

The telescope acts as a 'light bucket' for the shadow bands as they pass over it. The source of interest is usually isolated either with a focal plane diaphragm (optical and infrared observations) or with a pencil beam or beams (radio observations) in order to exclude as much extraneous radiation as possible.

Fig. 6.26 Theoretical occultation curves at 520 nm for different source diameters. (Published with permission from [301].)

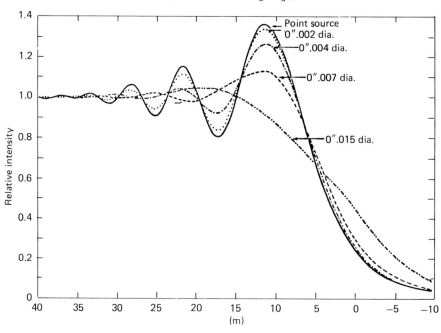

When the telescope diameter is a significant fraction of the fringe spacing the fringes will be blurred and their modulation reduced. Although this is not a problem at radio wavelengths because radio telescopes have a small aperture-to-wavelength ratio, occultation observations made with a 2 m optical telescope, for example, are limited to an angular resolution of about 0.001 arcsec through the smoothing effect of the telescope aperture. A broad bandpass leads to a similar effect as illustrated in Figure 6.28. For example, the angular resolution at 500 nm is also restricted to 0.001 arcsec by using a bandpass of 20 nm [301].

X-ray

Fringe detection is impractical for X-rays given the low flux of X-ray photons from cosmic sources and the brief dwell time of the fringes,

Fig. 6.27 Definition of the lunar motion vector. (Published with permission from [301].)

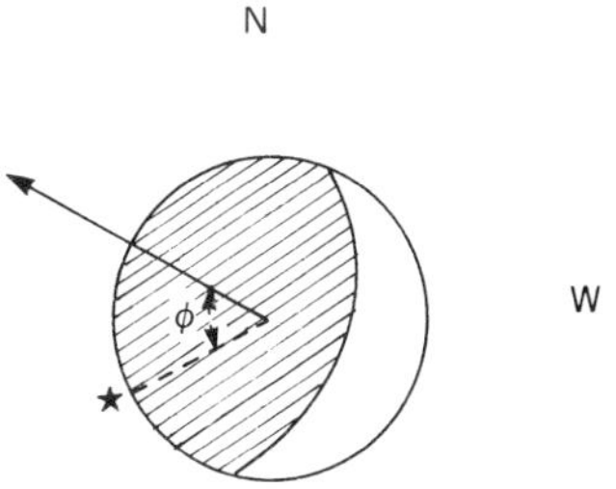

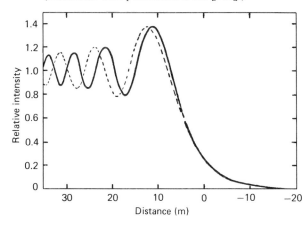

Fig. 6.28 Comparison of occultation fringe patterns at 450 nm and 550 nm. (Published with permission from [301].)

even assuming one could achieve a sufficiently narrow bandpass to avoid blurring.

Some limited mapping of the Crab Nebula was carried out, using balloon [325] and rocket [369] borne detectors during both the 1964 and 1974–5 series of lunar occultations. It was possible to separate the contributions of the point source (the pulsar) from the extended nebular component in the occultation curve and to demonstrate that the X-ray, non-thermal, radiation comes from a smaller area than the optical or the radio radiation.

Optical and infrared

According to equation (6.14) an optical occultation 'event' lasts less than a tenth of a second. The technique of estimating a source diameter from the occultation light curve is illustrated in Figure 6.29. A least squares fit is made to the data for a series of theoretical curves corresponding to a variety of source diameters [301]. The principal contributions to error in the fitting process are photon noise, scintillation, and irregularities at the lunar limb. The first two effects are a function of the telescope collecting area.

The effect of scintillation is well illustrated in Figure 6.29, which shows six occultations observed with either a 1.3 m or a 4 m telescope in the near infrared at 1.62 μm or 2.17 μm [328, 329]. Distance on the ground is given in metres or as milliseconds of arc at the lunar distance.

The distance on the ground indicated by the horizontal line with vertical bars lying above the data subtends the same angle at the Moon as the stellar disc which gives the best fit to the observations. The telescope diameter is shown by the horizontal line with the vertical marks immediately below the observations. Residuals in the sense (theoretical – observed) are plotted at the bottom. The oscillatory curve along the data zero line is the partial derivative of the theoretical curve with respect to the stellar diameter. The greatest amplitude of the curve occurs where it is most sensitive to the chosen stellar diameter.

Scintillation causes the low frequency noise component seen in the residuals for some of the data taken with the smaller telescope. As expected, there is no significant effect of scintillation with the larger telescope. Further, the 4 m data has a much higher signal-to-noise ratio. It appears that angular diameters as small as 1.5 to 2.5 milli-arcsec can be measured with a 4 m telescope despite the fringe blurring introduced by the large aperture.

Fig. 6.29 Occultation records for (a) W Tau, (b) γ Tau, (c) BD +15° 0635, (d) BS 2631, (e) IRC +20 168, (f) BD +13° 2045, obtained at Kitt Peak National Observatory with a 1.3 m telescope at 2.17 μm except for (b) which was obtained with a 4 m, and (c) and (e) which were made at 1.2 μm. Further details are given in the text. (Published with permission from [328].)

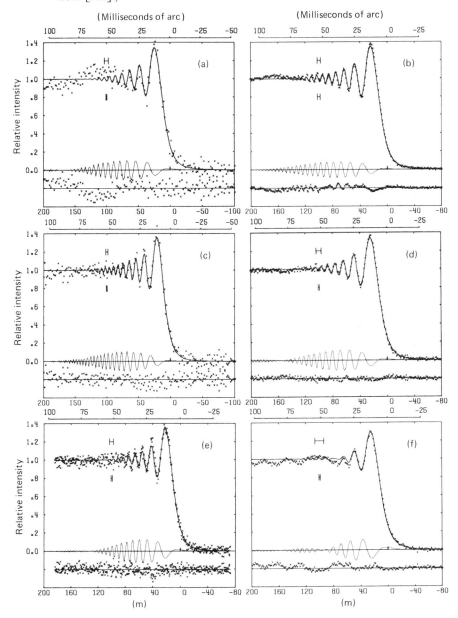

For most stars the beam size of the telescope at the Moon is little larger than its value on the ground; as a result, limb irregularities caused by boulders, craters, and mountain sides can introduce spurious variations into the fringe pattern. Independent, simultaneous observations from different sites or at different wavelengths from the same site should help to discriminate between effects of source structure and those due to local limb irregularities [124].

The star image is normally isolated by an aperture of 20 arcsec, or less, in the focal plane in order to reduce the contribution from the sky background and nearby stars [302]. Unless the telescope tracking and setting are highly stable and contain automatic corrections for differential refraction and telescope flexure, it is difficult to acquire the star in the centre of the aperture when it emerges from behind the Moon. As a result most stellar occultations are observed at immersion and they are made at the dark limb to avoid the simultaneous contamination by sunlight scattered at the bright limb. Thus occultation observations are normally made during the first and second quarters of the Moon.

Observations in the near infrared have the advantage that they can be made by day because of the relatively low level of the daytime sky brightness [329].

Telescopes in near-Earth orbit have velocities of the order of 7 km s^{-1}. As a result they will generally see the Moon in retrograde motion. When the Moon passes through the local meridian of the spacecraft there should be little apparent lunar motion. Such a slowing of the lunar motion coupled with the lack of an atmosphere to cause scintillation makes it possible to extend occultation observations to fainter stars in orbit than is possible from the ground [300].

Radio

The Moon is a thermal source while most cosmic radio sources of small angular size radiate non-thermally (see Chapter 1). As a result, at long wavelengths (i.e. greater than about 50 cm), the contribution of radiation from the Moon during an occultation observation is insignificant. Figure 6.30 shows the record of an occultation emersion made at 73 cm (410 MHz) of the radio source 3C273 [188]. One minute is indicated by the strong vertical lines. The occultation 'event' lasts of the order of a minute.

The 529 m long cylindrical telescope at Ootacamund (Ooty), India, was built specifically for the continuous monitoring of lunar occultations of radio sources at 326.5 MHz at a band width of 4 MHz [374, 377]. The telescope is built on a hillside inclined at the same angle to the horizontal as its geographical latitude ($+11° 22′ 50″$). As a result, the telescope is equatorially mounted and need only be rotated in hour angle in order to track the Moon. This form of telescope was chosen to achieve a large collecting area at modest cost.

A photograph of the telescope is shown in Figure 6.31, with a section of one of the 24 parabolic frames shown in Figure 6.32. The asymmetric design allows the dipole array at the focal line to be lowered to the ground for servicing. The reflecting surface is formed from 1100 stainless steel wires each of 0.38 mm diameter.

The array consists of 968 half-wave dipoles placed colinearly at 0.57λ intervals within a corner reflector (as shown in Figure 6.32). The dipoles are grouped into 22 modules of 44 dipoles each. The 22 dipoles on either side of the centre of a module are connected in series through 0.34λ continuously variable phase shifters placed between the dipoles. The signals from the 22 modules are combined to produce 12 simultaneous reception beams $3.6 \sec \delta$ minutes of arc apart (where δ is the declination) in the same way as the 81.5 MHz array at Cambridge discussed in Chapter

Fig. 6.30 A record of the occultation emersion of the radio source 3C 273. (Reproduced from [188].)

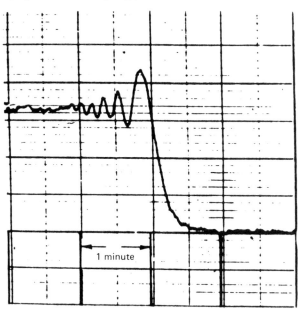

1 minute

Fig. 6.31 The 529 m long telescope at Ootacamund, India, built specifically for continuous monitoring of lunar occultations on a hillside which has the same inclination as its geographical latitude.

Fig. 6.32 A section of one of the parabolic frames of the Ooty telescope. (Reproduced from [377].)

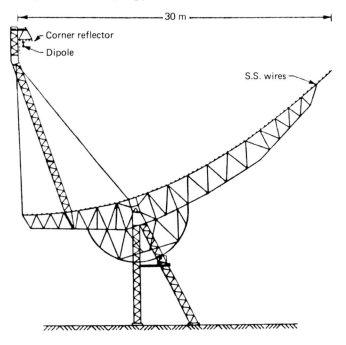

3. The 12 beams are operated simultaneously and are phase-switched to give cosine and sine outputs by correlating (multiplying) the 11 northern and the 11 southern modules. In right ascension the half-power beam-width is 2°.

The telescope tracks the Moon for 9.5 hours each day with the twelve beams exactly covering the lunar disc. Each beam can be moved 36 sec δ arc minutes either north or south and, as the Moon moves in hour angle, a beam centred on the Moon's outgoing edge is moved to the opposite edge. The technique of tracking by moving the beams from one edge to another is shown in Figure 6.33.

Although radio sources usually show a more complicated structure than a simple disc, the reduction of radio occultation data is, in principle, the same as that used at shorter wavelengths. The occultation curve of the source is simply the convolution of its angular brightness distribution normal to the edge of the Moon with the theoretical occultation curve for a point source. The brightness distribution normal to the lunar limb can then be recovered to the limiting resolution available in the fringe record by convolving the observed occultation curve with a suitable restoring function [347].

Fig. 6.33 Tracking of the Moon in declination by the twelve beams of the Ooty telescope. The beam centred on the outgoing lunar edge is switched to the opposite edge by appropriate phase switching of the 22 dipole outputs contributing to each beam. (Reproduced from [377].)

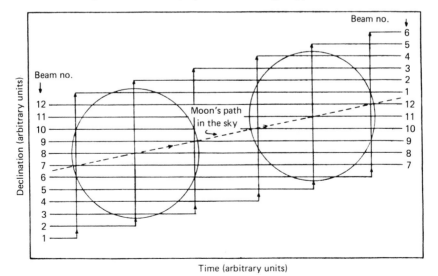

Time (arbitrary units)

7

Single channelled detectors

The distinction between coherent and incoherent detection was drawn in Chapter 2. The signal amplitude from a coherent, or thermal, detector, is proportional to the incident *energetic* flux, while that from an incoherent detector (excluding thermal), is proportional to the incident *photon* flux.

Incoherent detection

The performance of incoherent detectors at wavelengths less than 1 μm is best described in terms of responsive quantum efficiency (RQE), detective quantum efficiency (DQE) [222], and, in the case of multi-channelled detectors, by the modulation transfer function (MTF) [110].

Responsive quantum efficiency (RQE)

Often called the quantum efficiency, q, the responsive quantum efficiency is that fraction of the photons incident on the detector which contribute to the output signal,

$$q = N'/N \tag{7.1}$$

where N is the total number of incident photons of which N' are absorbed and contribute to the output signal. q varies with wavelength (photon energy), and, in the case of nonlinear detectors such as photographic emulsions, it is also a function of N.

The quantum efficiencies of various commonly used photocathode materials are shown as a function of wavelength in Figure 7.1. The photocathodes are either semitransparent which means that light is incident on one side and the photoelectrons are drawn off on the other, or they are opaque in which case the photoelectrons emerge on the same side as the incident light.

The long wavelength at which the sensitivity begins to fall off corresponds to the excitation energy of the photoelectrons or carriers. The abruptness of the fall in sensitivity at longer wavelengths is proportional to the phonon density and hence the temperature. The lower the temperature the sharper the long wavelength cutoff. There is also a tendency for the work function of photocathode materials to increase with decreasing temperature thereby decreasing the quantum efficiency (the

Fig. 7.1 The spectral response of some commonly used photocathode materials in the optical and the ultraviolet. The (responsive) quantum efficiencies are shown. The S-devices and those marked ST are semi-transparent.

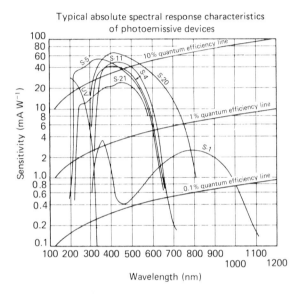

Typical absolute spectral response characteristics of photoemissive devices

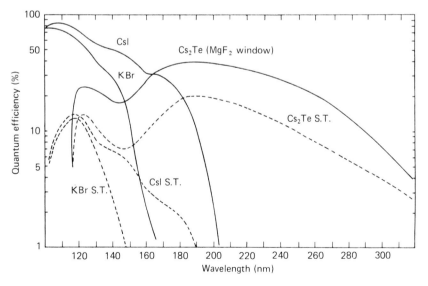

work function being the average energy lost by a photoelectron before it emerges from the photocathode layer).

RQE enhancement

Sensitivity to UV radiation can be enhanced for a 'visible' photocathode by down conversion of UV photon energies and their associated frequencies to the visible using a thin layer of phosphor material on the photocathode window. Sodium salicylate is the most widely used phosphor [345]. It is uniformly efficient between 30 and 250 nm and reaches a peak conversion efficiency of about 60 per cent for a 10 μm thick layer. The peak in the fluorescent spectrum is at 420 nm with a width of about 80 nm. Hence it is ideal for blue sensitive photocathodes.

The organic phosphor coronene, $C_{14}H_{12}$ (sometimes called mouse milk), has a constant conversion efficiency between 150 and 350 nm, which is higher than that for sodium salicylate. The peak of the fluorescence is at 500 nm which is more appropriate, although not optimal, for silicon detectors [212]. The CCD detectors of the Wide Field Planetary Camera of the Hubble Space Telescope have a 5 μm layer of coronene on them [18]. The question of the UV sensitivity of CCDs and quantum efficiency hysteresis are discussed in Chapter 8.

Total internal reflection can be used to enhance the quantum efficiency of semi-transparent photocathodes [117].

Detective quantum efficiency (DQE) [222]

The maximum information available in a beam of N photons is given by the signal-to-noise, $N^{0.5}$, as discussed in Chapter 2. The incomplete detection implied by the quantum efficiency, q, together with additional noise introduced by the detection system, further limits the available information. The detective quantum efficiency, DQE, is a measure of the ability of the detector to reproduce the information available in the original beam and it is defined as

$$DQE = ((s/n)_{out}/(s/n)_{in})^2 \tag{7.2}$$

where (s/n) is the signal-to-noise ratio, or, more usefully,

$$(s/n)_{out} = (s/n)_{in}(DQE)^{0.5} \tag{7.3}$$

There are two principal photon detection techniques:
(a) analogue integration,
(b) photon counting.

The principles of operation for signal generating detectors are shown for single channelled detectors in Figure 7.2 and they are readily generalised to multi-channelled devices which are discussed in Chapter 8. The DQE of photographic emulsions which provide a permanent record of each exposure rather than an electronic signal is also discussed in Chapter 8.

Analogue integration

The detectors used in this mode generally have little or no internal gain. Photoelectrons are allowed to accumulate in the detector during an exposure. At the end of the exposure the accumulated photoelectrons are sensed as a voltage or current through an impedance matching amplifier. The amplifier and the detector introduce a noise contribution ($D^{0.5}$) which is equivalent to the shot noise from an additional D photoelectrons. (Detector, or system, noise is generally quoted as $D^{0.5}$.) In this case

$$\text{DQE}_{analogue} = q/(1 + (D/Nqm^2)) \tag{7.4}$$

or

$$(s/n)_{out} = (Nq/(1 + (D/Nqm^2)))^{0.5} \tag{7.5}$$

where N photons are incident on the detector during the exposure and, if the detector is multi-channelled, m is the MTF of the detector at the spatial frequency of interest or is a mean over a range of spatial frequencies. For a single channelled detector m is taken as unity. Strictly speaking, equations (7.4) and (7.5) should be multiplied by a factor, $s(N, N_0)$, which takes account of detector saturation, where N_0 is the detector storage capacity. The form of $s(N, N_0)$ depends on the detector and is discussed later in this chapter.

Fig. 7.2 The principles of single channel integrating and photon counting detectors.

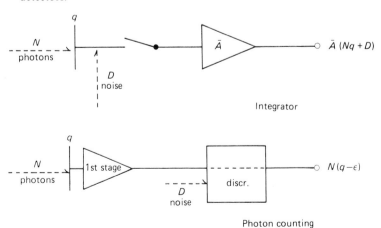

Photon counting

In this mode the internal gain of the detector is so large that the signal amplitudes associated with individual photoelectrons exceed the signal fluctuations associated with the majority of the noise sources. During an exposure the detector output is connected to a discriminator (as shown in Figure 7.2) where the threshold is adjusted to pass the majority of the photoelectron pulses while blocking the majority of the pulses caused by the noise. The output of the discriminator is fed to a counter. The detective quantum efficiency is given by the general relation

$$\mathrm{DQE}_{\text{photon counting}} = ((q-\varepsilon)g(N, n, t_0))/(1 + an/N(q-\varepsilon)m^2) \qquad (7.6)$$

or

$$(s/n)_{\text{out}} = (N \cdot \mathrm{DQE})^{0.5} \qquad (7.7)$$

where N photons are incident per second on the detector and there are n noise pulses per second. ε is the fractional loss of photon events in the discriminator, a is the fraction of noise pulses passed by the discriminator, and the factor $g(N, n, t_0)$ compensates for coincidence losses in the counting.

Pile-up errors

Each detected photon or noise pulse has a finite duration, t_0, which introduces a dead time in the detector (case A) or in the detection circuit (case B). As a result, the actual exposure time or the number of pulses detected is reduced by the degree of overlapping of dead times. If the pulse arrival times can be described by a Poisson distribution (equation (2.11)) then the factor $g(N, n, t_0)$ is given by

$$g(N, n, t_0) = \exp(-(N(q-\varepsilon) + an)t_0) \qquad (7.8a)$$

or one can define the *pile-up error* as $1 - g$.

Other approximate formulae used to correct for pile-up errors are

$$N = n/(1 - t_0 n) \qquad (7.8b)$$

or

$$N = n/(1 - an^{(1+b)}) \qquad (7.8c)$$

where N is the true rate, n the apparent rate, and $t_0 = an^b$ [134].

The silicon diode: an analogue detector

Figure 7.3 shows a circuit schematic for a silicon diode detector. The diode is formed from single layers of p- and n-type silicon overcoated with silicon dioxide. In operation the diode is reverse biased by some 5 volts (n positive relative to p). This creates a depletion region of about

$2 \, \mu m$ width at the interface between the layers. The diode is disconnected from the bias before an exposure by opening the switch. Incoming photons generate charge carrier pairs (one electron and one hole per absorbed optical photon) which separate in the depletion region and partially discharge the bias. The diffusion length for the minority carriers is between 25 and 50 μm.

At the end of the exposure the diode is rebiased by closing the switch. The charge necessary to rebias the diode is sensed and digitised by the readout circuitry. The rebiasing charge is directly proportional to the number of charge carriers generated during the exposure and, consequently, it is directly proportional to the number of absorbed photons.

Reset (KTC) noise

Although Johnson noise associated with the amplifier and any resistors is important, reset noise is fundamental to this type of device [151, 405]. When the bias voltage is applied to the diode through the switch, Johnson noise in the switch resistance introduces an uncertainty in the actual value of the bias. Provided the switch is closed for a time which is long compared with the circuit settling time, the r.m.s. variation, ΔV, between successive bias voltages is given by

$$\Delta V^2 = kT/C \quad \text{V} \tag{7.9}$$

where T is the absolute temperature, C is the stray capacitance, and k is Boltzmann's constant. Alternatively, the r.m.s. fluctuation, ΔQ, in a succession of bias charges is given by

$$\Delta Q^2 = kTC \quad \text{C}^2 \tag{7.10}$$

Fig. 7.3 Principle of operation of a silicon diode detector.

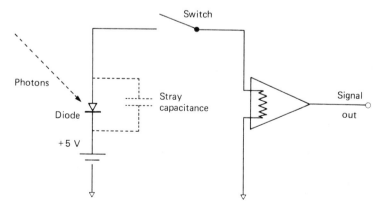

or

$$\Delta Q = 2.5 \times 10^9 (kTC)^{0.5} \quad \text{electrons} \tag{7.11}$$

C for a single diode with a carefully designed output circuit is normally less than 1 pF ($= 1$ picofarad $= 10^{-12}$ farad). If the diode is maintained at a temperature of 150 K (see Figure 2.8) then, from equation (7.11), the reset noise, ΔQ, is some 280 electrons. This noise is equivalent to that expected from $(280)^2$ detected photons, i.e. D would be 78 400 in equations (7.4) and (7.5).

Dark noise

Charge carriers are produced thermally which leads to the equivalent of a leakage current. For semiconductors with wide energy gaps such as silicon the dark leakage current, $J(T)$, is caused almost entirely by the production of hole–electron pairs in the space charge region [416]. If surface leakage currents are negligible then the temperature dependence of the leakage current is given by [342]

$$J(T) = 3.11 \times 10^3 (wa/t) T^{1.5} \exp(-7015/T) \quad \text{A} \tag{7.12}$$

or

$$j(T) = 1.94 \times 10^{22} (wa/t) T^{1.5} \exp(-7015/T) \quad \text{electrons s}^{-1} \tag{7.13}$$

where w is the width of the depletion region, a is the area of the junction, and t is the minority carrier lifetime. Equation (7.12) implies that the dark current approximately doubles for every 8 K rise in temperature and would saturate in about one second at room temperature. Empirical values tend to be somewhat smaller [152].

The theoretical relation between temperature and dark current for a multi-channel silicon detector (Reticon) is shown in Figure 7.4 together with points measured at different temperatures [70]. If the temperature is sufficiently stable then the dark current can be removed by taking exposures with a dark slide closed from time to time.

Unfortunately, as can be seen in equations (7.12) and (7.13), the magnitude of the leakage current is directly proportional to the width of the depletion region, ω, and hence to the bias remaining on the diode. This means that the appropriate dark current is also a function of the number of photons detected during an exposure. This awkward problem is circumvented by maintaining the diode at a low enough temperature that the dark current contribution is negligible.

Taking the values in the above example, the dark current would introduce a noise component $(j(T)\tau)^{0.5}$, where τ is the exposure time in seconds and the variation of DQE with N for this diode ($T = 150$ K,

$C = 1\,\mathrm{pF}$) at 800 nm where its responsive quantum efficiency, q, is 0.8, would be

$$\mathrm{DQE} = 0.8/(1 + (78\,400 + (j(T)\tau))/0.8N) \tag{7.14}$$

or

$$(\mathrm{s/n})_{\mathrm{out}}/(\mathrm{s/n})_{\mathrm{in}} = (\mathrm{DQE})^{0.5} \tag{7.15}$$

The DQE and $(\mathrm{s/n})_{\mathrm{out}}/(\mathrm{s/n})_{\mathrm{in}}$ are plotted as a function of Nq, the number of detected photons, in Figure 7.5 for exposures which are short enough that the dark current can be ignored. A saturation charge of 2×10^7 electrons has been assumed which leads to the cutoff at high N.

The photomultiplier: a photon counting detector [431]

Figure 7.6 shows, in outline, the structure of a number of different photomultipliers. Photoelectrons released by photons absorbed in the photocathode are accelerated through a potential of some 100 V to the

Fig. 7.4 Dark current as a function of temperature for a single silicon diode. (Published with permission from [70].)

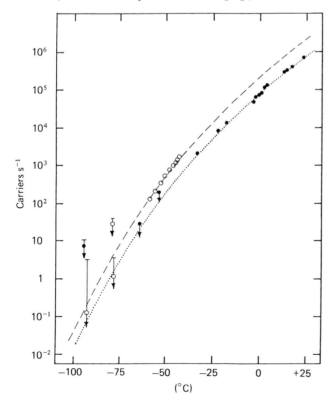

Carriers s^{-1}

$(^\circ\mathrm{C})$

first dynode where they release secondary electrons. The secondary electrons in turn are accelerated towards a second dynode where they are again multiplied. Most photomultipliers have some ten dynodes. The gain, g, at each dynode is roughly proportional to the energy of the impinging electrons, i.e. $g = AV$, approximately (where V is the accelerating voltage and A is a constant). The output from the last dynode is collected by the anode to give a current pulse. The overall gain, G, of the photomultiplier usually lies between 10^5 and 10^6 and is simply g^p, where p is the number of dynodes.

The gain stability depends on the voltage stability approximately as

$$dG/G = p(dV)/V \qquad (7.16)$$

since p is of order 10 then, in order to achieve an overall gain stability of, say, 0.1 per cent, the high voltage fluctuation must be < 0.01 per cent.

The anode output can be integrated on a capacitor or simply displayed on a chart recorder [403]. In this case the device acts as an analogue detector. However, the great advantage of a photomultiplier is its ability to detect individual photoelectrons at very low light levels. For good photon counting characteristics it is important that the probability distribution of the dynode gain be Poisson. If the mean gain per dynode is g, then the number of secondary electrons may be any integer, k, which

Fig. 7.5 The variation of DQE and $(s/n)_{out}/(s/n)_{in}$ with the number of detected photons, Nq, for a silicon diode.

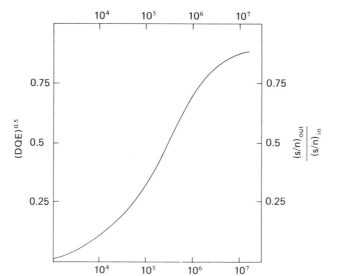

Detected photons Nq

occurs with a probability $P(g, k)$ given by

$$P(g, k) = ((g^k)/k!) \exp(-g) \qquad (7.17)$$

The r.m.s. deviation of this distribution is $g^{0.5}$. For most photomultipliers g lies between 2 and 6. An important property of this distribution is the significant probability of observing zero secondaries from the first dynode (lost photoelectrons). If $g = 2$ the predicted loss is 14 per cent, for $g = 3$ the loss is 5 per cent, and if $g = 4$ the loss is 2 per cent, with a corresponding reduction in q [431]. This emphasises the importance of having a high gain at the first dynode. It can also be shown that the pulse height

Fig. 7.6 The dynode structure of five different photomultipliers.

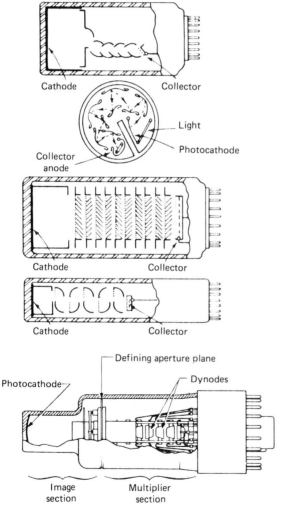

distribution at the anode is largely determined by the values of g for the first few dynodes.

The finite widths of individual pulses arise from the slightly different paths followed by the electrons through the multiplier structure, leading to a spread in arrival times at the anode. The pulse will be further broadened by the time constant associated with stray capacitance in the output circuitry and the load resistor. The latter is typically about $50\,\Omega$ for pulse counting. Typical full width half maximum (FWHM) pulse widths for photon counting photomultipliers lie between 5 and 20 ns.

Dark noise

The principal sources of photomultiplier noise are:
  (i) thermally excited electrons emitted from the photocathode and dynodes,
 (ii) electrons released from the dynodes by positive ions from residual gas in the photomultiplier envelope,
(iii) cosmic-ray-induced Čerenkov radiation in the preceding optics,
(iv) scintillations from electrons colliding with the glass envelope of the photomultiplier,
 (v) leakage currents from the dynode pins to the anode,
(vi) glow from the final dynodes and anode.

The contribution of each of these noise sources can be minimised as follows:

(i) Photocathode thermal emission is reduced to an acceptable level by refrigerating the photomultiplier in accordance with the curve in Figure 2.8. The thermal emission is further reduced if the area of the photocathode is limited to about $1\,mm^2$ by the use of a diaphragm in front of the first dynode and by focussing the electrons with an axial magnetic field. Pulses from thermal electrons released from the dynodes are generally of smaller amplitude than those originating from the photocathode and they are excluded by the discriminator.

(ii) Most photocathode materials will not survive the standard high temperature bake-out necessary to create a good vacuum, and in consequence a significant quantity of gas is often left absorbed in the tube. When the high voltage is applied ions from this gas travel towards the cathode releasing bursts of electrons at each dynode which eventually cause giant pulses at the anode. The number of ions can be minimised by storing photomultipliers in the dark even when not connected to the high voltage. To maintain a really low dark current the photomultiplier should be kept continuously refrigerated and under high voltage. Most photomultiplier manufacturers will, for a price, select photomultipliers with lower than average dark current.

(iii) This effect is strongly dependent on the altitude of the observatory. Spectrally, the number of photons from Čerenkov radiation increases linearly with frequency or as λ^{-2} (see Chapter 1). Consequently, the amount of radiation admitted increases rapidly as the bandpass is extended into the UV. Quartz and UV transmitting glasses transmit a much wider band than regular glasses, particularly the vacuum UV radiation which, for photocathodes with excitation energies appropriate to the visible spectrum, will also tend to produce giant pulses because of photoelectron pair production. Hence, quartz and UV transmitting glasses should be avoided in the optics, particularly for the photomultiplier window, unless good UV transmission is essential. Envelope glasses containing radioactive material should also be avoided.

The effect of charged cosmic ray particles can be limited by surrounding the photometer with Geiger counters and gating the counter off whenever one of them detects a charged particle.

(iv) While the photomultiplier can be operated with either the cathode or the anode at ground potential, the anode is normally kept close to ground for safety. In this case there is a danger that corona will develop between the cathode and those parts of the surroundings which are at ground. This is best avoided by keeping the photomultiplier in a vacuum or a desiccated atmosphere. The photomultiplier must be surrounded by a shield at cathode potential otherwise the glass envelope tends to assume ground potential, thereby attracting the electrons within the tube to the wall where they cause scintillations. Magnetic shielding is also essential.

(v) The resistance between the anode and the other dynode electrodes should be of the order of $10^{14}\,\Omega$ to avoid significant leakage current from the dynodes to the anode. This is usually achieved by surrounding the anode with a guard ring at either ground or anode potential.

(vi) At high enough photocurrents the final dynodes and the anode may glow. The light may be ducted back to the photocathode through the envelope. The effect can be eliminated by proper photomultiplier design.

Figures 7.7 and 7.8 summarise the essential adjustments for optimal sensitivity in a photon counting system. Figure 7.7 shows a schematic pulse height distribution with, and without, light incident on the photocathode. With no light the dark current appears as two exponential distributions, one arising from thermal noise at the dynodes and the photocathode, and the other from the large amplitude pulses caused by cosmic rays and ions. The peak, in the presence of light, corresponds to photoelectrons from the photocathode. The two vertical lines indicate the approximate settings for the discriminator 'window' which maximises the signal-to-noise for light induced pulses.

Figure 7.8 indicates how counting rates in the dark and with light vary with applied voltage. Ideally the count rates will increase until a plateau is reached indicating that all of the cathode electrons are being counted (i.e. $\varepsilon = 0$ in equation (7.6)). Further increase in the voltage tends to increase the dark current at the expense of the cathode pulses. This departure of the dark current from the ideal is probably due to point discharges from irregularities in the dynode structure.

DQE of a photon counting photomultiplier system

Figure 7.9 shows the variation of $DQE^{0.5}$ and $(s/n)_{out}$ with N for a photon counting system having a photomultiplier with an S-20

Fig. 7.7 A schematic pulse height distribution for a photomultiplier in the presence and absence of light incident on the photocathode.

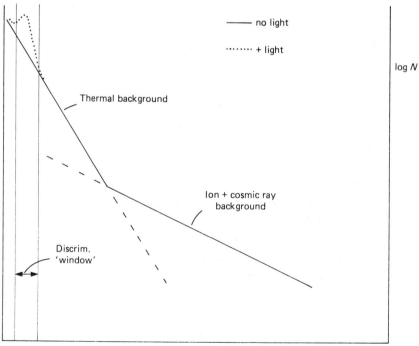

Fig. 7.8 The variation of photomultiplier counting rate with applied voltage in the presence and absence of light. Optimum voltage adjustment is in the plateau region at the maximum signal/dark-noise ratio as shown.

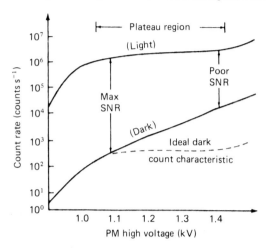

Fig. 7.9 The variation of $DQE^{0.5}$ and $(s/n)_{out}$ with the number of incident photons for a photon counting system using a photomultiplier with an S-20 photocathode and a photocathode dark current of 5 photoelectrons per second.

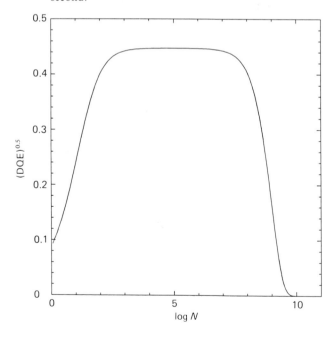

photocathode. Using the values $q = 0.2$ ($\lambda = 420$ nm), $t_0 = 10$ ns, $\varepsilon = 0$, and $an = 5$ gives, from equations (7.6) and (7.7),

$$DQE = 0.2g/(1 + 25/N) \qquad (7.18)$$

where $g = \exp(-2N \cdot 10^{-9})$ if one ignores the value of (an) for high light levels (see equation (7.8a)). The falloff at high light levels is caused by the 'pile-up' or coincidence losses discussed earlier, where $(1 - g)$ is the pile-up error. At a high enough light level the system will block completely.

A photoelectric photometer

Figure 7.10 shows, schematically, the principal components of a single channel photometer. The Fabry lens focusses an image of the telescope exit pupil onto the photocathode. Use of the Fabry image eliminates errors caused by motion of the star image in the entrance diaphragm coupled with point-to-point variations in sensitivity over the surface of the photocathode. The exit pupil illumination does not change for small motions of the stellar image unless the edge of the diaphragm 'knife-edges' it.

The filters are shown in front of the diaphragm where there is more room and for ease of interchange but could also have been placed behind it. In more general applications the filters may be replaced by a monochromator.

Typically there are two possible circuits which might be associated with the photomultiplier anode.

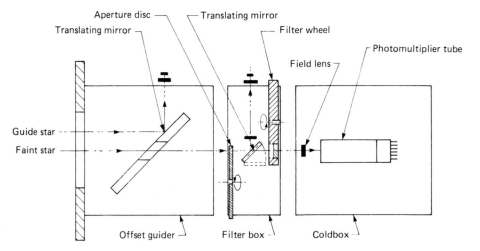

Fig. 7.10 Schematic of a modular, single channel photometer.

(i) Integrating circuit [403]. The photomultiplier acts as a current source with the capacitor integrating charge during an exposure.

(ii) Pulse sensing and shaping circuit which delivers a standard pulse to the counter for each signal pulse passed by the voltage discriminator.

Photometric observations must be corrected for the effect of differential extinction in the Earth's atmosphere. The photometric precision is improved if measurements through the various filters can be made simultaneously or in sufficiently quick succession before the sky transparency changes significantly. In consequence some photometers are multi-channelled [308, 403].

Photometric observations [179, 431]

A full discussion of photometric data reduction is beyond the scope of this book. For this, the reader should refer to one of the primary references. What follows is an outline of principles.

The colours and magnitudes of programme objects are compared with those of standard stars. The bandpasses of the photometric system are defined by the convolution of the transmissions of the monochromator, telescope, and atmosphere with the quantum efficiency of the detector. These bandpasses are usually arranged to be as close to those of the original standard system as possible. The effective wavelength of the detected radiation (i.e. λ weighted according to detected intensity) is a function of the photon spectral distribution and the chromatic, air mass dependent portion of the atmospheric transmission. These chromatic effects are important for broad band photometry of the UBV type but not for quasi-monochromatic filters of a few nanometres width. Finally, corrections must be made for contamination by out-of-band radiation (discussed below).

The standard stars should cover a greater range in colour than that expected for the programme objects. At least two standard stars, covering the full range of colour, should be observed through the same or greater range of air mass as the programme objects in order to provide accurate chromatic values of zenith extinction (see Chapter 2).

Monochromators

With the continual improvement in detective quantum efficiency of area detectors such as the Charge Coupled Devices (see Chapter 8), wide band, coloured glass filters tend to be replaced by narrower band interference filters. For single or few channel photometers, spectrometers (see Chapter 5) are often preferred as monochromators [308, 403].

Absorption filters

These absorb or scatter transmitted light generally through ionic absorption. Coloured glasses have been known throughout history and they have been supplemented more recently by gelatin, liquid, plastic, and gases. The filters are divided into two types.

(a) Blocking, cut-on, sharp cut, or temperature coloured, which are highly transmitting beyond a certain long wavelength and suppress shorter wavelengths.

(b) Sine or bell-curve, which transmit a band of radiation. They have the characteristics of a cut-on filter at the short wavelength limit and a more gradual drop in transmission on the long wavelength side.

Class (b) filters usually have a long wavelength 'leak' which has to be either calibrated or blocked with another filter. One class (b) filter which does not suffer from this problem is copper sulphate. The importance of long wavelength leaks is discussed below.

The fraction of light transmitted by a filter is a function of the reflection losses, $R(\lambda)$, and the absorption coefficient, $\alpha(\lambda)$, both of which are wavelength dependent. The transmission at normal incidence of an air/glass interface, T, is

$$T = 4nn_0/(n_0 - n)^2 \tag{7.19}$$

where n and n_0 are the refractive indices of the glass and air ($=1$), respectively. The transmission of an absorbing filter is

$$T(\lambda) = I(\lambda)/I_0(\lambda) = \rho \exp(-\alpha\lambda x) \tag{7.20}$$

where ρ is the correction factor for reflection losses and x is the filter thickness, I is the transmitted intensity and I_0 is the incident intensity. Sometimes manufacturers specify the internal transmittance, $\tau(l)$, for some standard thickness l,

$$\tau(l) = \exp(-\alpha l) \tag{7.21}$$

in this case the transmission formula becomes

$$T(\lambda) = \rho\tau^{(L/l)} \tag{7.22}$$

where L is the actual thickness of the filter.

At normal incidence the reflection loss is given by

$$R(\lambda) = ((n - n_0)/(n + n_0))^2 \tag{7.23}$$

and

$$\rho = (1 - R)^2 \qquad (7.24)$$

Absorption increases as the angle of incidence is increased because the effective path length in the filter increases by the factor $(1 - (\sin^2 \theta)/(n^2))^{-0.5}$.

Reflection also increases and the effect is a function of polarisation, but the amount is small until the critical angle is approached.

The use of absorption filters of the type described above is largely restricted to the optical spectrum.

Interference filters [303]

These are based on the principle of the Fabry–Perot interferometer (see Chapter 5). When collimated light is incident at an angle θ on two parallel reflecting surfaces separated by a transmitting medium of refractive index n and thickness d, all wavelengths, λ, are preferentially transmitted if the spacer satisfies the half-wave plate criterion

$$2nd \cos \theta = m\lambda \quad \text{(maxima)} \qquad (7.25)$$

where m is an integer. Interference filters normally operate in the zero order (i.e. $\cos \theta = 1$).

Metallic films tend to absorb some of the transmitted light. In the *all-dielectric* filter the reflecting layers are made of non-absorbing dielectric materials of different refractive indices deposited as high reflectance $\lambda/4$ coatings enclosing a $\lambda/2$ spacer. The *induced transmission* filter uses $\lambda/4$ high reflectance dielectric layers enclosing a $\lambda/2$ spacer between metallic reflectors.

The half-width of the bandpass is a function of the finesse, F (equation (5.20)), and the interference order m. For both types of interference filter the bandpass can be sharpened by increasing the number of high reflectance layers and by coupling several identical filters. For the simple Fabry–Perot etalon $T(\lambda)$ has the form

$$T(\lambda) = (1 + F \sin^2 \delta/2)^{-1} \qquad (7.26)$$

where δ is the phase as defined in equation (5.16).

Absorptive blocking filters are necessary to exclude harmonic transmissions at wavelengths other than the one of interest. Figure 7.11 shows typical transmission characteristics of both all-dielectric and induced transmission filters together with an appropriate blocking filter.

The wavelength of peak transmission of an interference filter is sensitive to the angle of incidence and, where the reflecting films are metallic, to the polarisation of the incident light. The change in peak wavelength is given by

$$\lambda_1/\lambda_0 = (n_0^2 - \sin^2 \theta)^{0.5}/n_0 \tag{7.27}$$

where λ_1 is the wavelength of peak transmission at the angle of incidence θ and λ_0 is the wavelength of peak transmission at normal incidence. n_0 is the effective refractive index of the whole filter,

$$n_0 = n_1/(1 - n_1 n_2 + (n_1/n_2)^2)^{0.5} \tag{7.28}$$

where n_1 and n_2 refer to the low and high indices of refraction, respectively.

Although they provide well defined controllable bandpasses,

Fig. 7.11 Typical transmission characteristics of (a) an all-dielectric filter and (b) an induced transmission filter. (Reproduced from [303].)

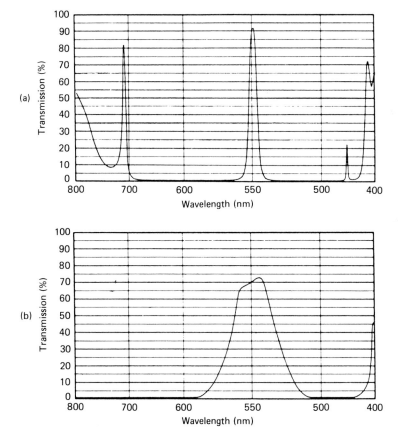

interference filters still have some drawbacks compared with absorption filters. The displacement of the peak transmission to shorter wavelength with increasing angle of incidence limits the speed (focal ratio) of the optics which feed the filters (especially for narrow band filters) if serious systematic variations of peak λ across the filter are to be avoided. It also means that filter tilt must be highly repeatable to avoid systematic photometric errors.

The peak transmission for broad band filters is about 60 per cent in the visible but more like 20 per cent in the UV. In general the narrower the filter bandpass the lower is the peak transmission. Further, although well peaked, most interference filter transmissions have extensive transmission wings, particularly on the long wavelength side.

Photometric errors from out-of-band radiation

There is a wide range in stellar temperatures and the photon spectral distributions are distorted by strong absorption features (see Chapter 1). When a filter transmission is convolved with the spectral energy distribution, the detector response, and the transmission of any other optical components in the telescope beam (including the atmosphere), the contribution of out-of-band radiation may be a significant or even a dominant part of the total signal from the detector.

Figure 7.12 shows a simplified transmission curve for a good quality UV filter. The log of the ratio $T(\lambda)/T_p$ is plotted against λ, where $T(\lambda)$ is the transmission at wavelength λ and T_p is the peak transmission of the filter. The bandpass is 10 nm, centred on 150 nm, and the transmission is constant, T_p, within the bandpass. At 155 nm, $T(\lambda)/T_p = 0.1$. The transmission falls off exponentially on the long wavelength side and is zero on the short wavelength side. The exponential tail is the critical factor in the analysis and can be characterised by a rejection ratio $R = T(555)/T_p$. A value of $R = 10^{-4}$ is used here, which is a reasonable value for a good quality filter.

The output of incoherent detectors is proportional to the number of detected photons. (This is not quite true for wide band detectors such as UV sensitive CCDs where a UV photon produces 2 or 3 carrier pairs compared with one pair in the visible – a weighting factor can be readily incorporated into the analysis.) In Figure 7.13(a, b, c, d, and e) the photon spectra, $n(\lambda)$, for thermal sources with temperatures of 3.5×10^4, 2×10^4, 10^4, 5×10^3, and 3.5×10^3, respectively, are shown by dashed lines. The

Fig. 7.12 The simplified log transmission curve of the UV filter used in the discussion of out-of-band radiation. From a diagram by R. W. O'Connell.

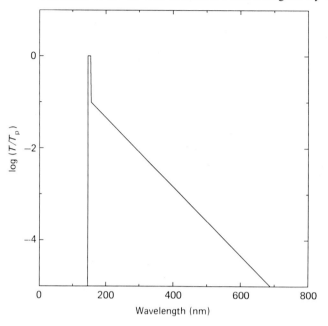

Fig. 7.13 The photon spectra (dotted lines) and the normalised product of the photon spectra and the transmission curve of Figure 7.12 (solid lines) for temperatures of: (a) 3.5×10^4 K, (b) 2×10^4 K, (c) 10^4 K, (d) 5×10^3 K, and (e) 3.5×10^3 K.

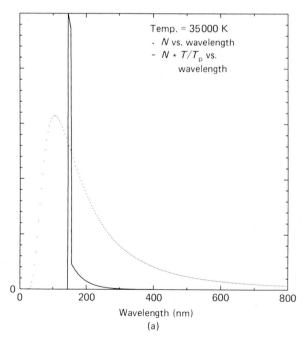

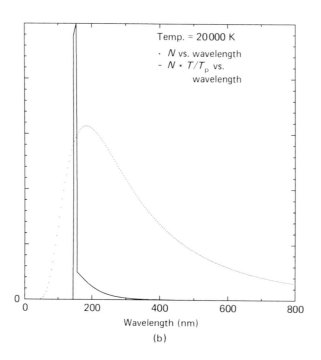

(b)

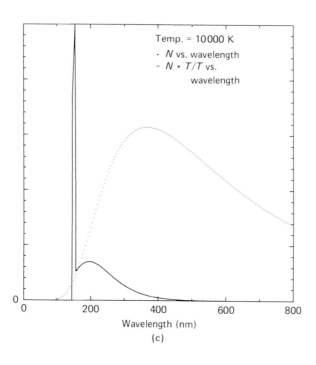

(c)

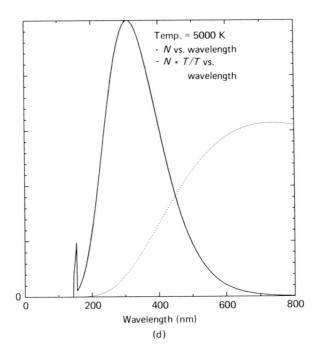

Temp. = 5000 K
· N vs. wavelength
- N * T/T vs.
    wavelength

Wavelength (nm)

(d)

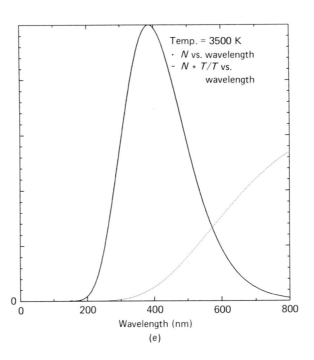

Temp. = 3500 K
· N vs. wavelength
- N * T/T vs.
    wavelength

Wavelength (nm)

(e)

solid lines, which are normalised to their maximum values, are the product of the transmission curve of Figure 7.12 and the photon spectra.

While the thermal curves do not closely mimic actual stellar photon spectra, especially at the lower temperatures, the transmitted photon spectra shown by the solid lines graphically illustrate the problem of the 'red' leak for a broad band detector.

(i) Only in the case of the 'O stars' at 3.5×10^4 and 2×10^4 K is the out-of-band radiation less than that transmitted within the band.

(ii) The principal wavelength region contributing to the out-of-band radiation moves to longer wavelengths as the temperature is lowered. Consequently, for accurate photometry of cool objects measurements in the 150 nm band must be accompanied by measurements through a series of filters at progressively longer wavelengths.

(iii) For cool objects, photon noise from the out-of-band radiation will dominate the small signal from the band itself.

(iv) The importance of the out-of-band radiation increases inversely as the width of the filter bandpass.

Ideally these problems can be solved if the detector has an excitation energy just less than the long wavelength edge of the filter. 'Solar blind' detectors which have no visible sensitivity are a considerable improvement but there would remain the 'UV leak' problem for the hotter stars.

O stars are rare and generally occur in crowded fields in which the majority of the other stars are cool, especially when detected in other galaxies. Blue subdwarfs and white drawfs are often members of close binaries where the other star is much cooler.

Specification of a filter rejection factor requires details of the quantum efficiency of the detector and the introduction of actual stellar photon spectra. While no firm figures can be suggested, a value of $R = 10^{-6}$ would seem to be necessary for the latter studies with a silicon detector.

The red leak problem is important in ground based ultraviolet measurements but is less acute.

Polarimetry [357, 358]

With the addition of a polarisation analyser a photometer can be transformed into a polarimeter. Polarisation is described in terms of the orientation of the electric vector [90]. A monochromatic wave propagating in the z direction can be described by its components of vibration in the orthogonal planes xz and yz,

$$E(x) = A(x) \exp(i(\omega t - 2\pi z/\lambda + \delta(x))) \tag{7.29}$$

$$E(y) = A(y) \exp(i(\omega t - 2\pi z/\lambda + \delta(y))) \tag{7.30}$$

where $A(x)$ and $A(y)$ are the amplitudes of the x and y vibrations, and $(\delta(y) - \delta(x))$ is the difference in phase between the two components.

Stokes parameters provide a full description of the state of polarisation of the wave and they are defined as

$$I = \overline{A(x)^2} + \overline{A(y)^2} \tag{7.31}$$

$$Q = \overline{A(x)^2} - \overline{A(y)^2} \tag{7.32}$$

$$U = \overline{2A(x)A(y)\cos(\delta(y) - \delta(x))} \tag{7.33}$$

$$V = \overline{2A(x)A(y)\sin(\delta(y) - \delta(x))} \tag{7.34}$$

where overbars indicate time average.

The degree of linear polarisation, p, is given by

$$p = (Q^2 + U^2)^{0.5}/I \tag{7.35}$$

or, when the angle of polarisation is known,

$$P = (I_1 - I_2)/(I_1 + I_2) \tag{7.36}$$

where I_1 and I_2 are the intensities measured at the angle of polarisation and perpendicular to it, respectively.

The direction of the maximum of polarisation, θ, is given formally by

$$\tan 2\theta = U/Q \tag{7.37}$$

The values of U and of Q depend on the choice of axes. The usual reference axis for astronomical observations is declination for x and right ascension for y with θ increasing from $0°$ through $180°$ going from north through east.

The degree of circular polarisation, q, is given by

$$q = V/I \tag{7.38}$$

V describes the shape of the polarisation ellipse and the sense of rotation of the electric vector. For positive, circular polarisation, θ increases with time. While there is some confusion over the definitions of handedness, the most appropriate defines right-handed circular polarisation as positive.

Astronomical sources exhibit both linear and circular polarisation (see Chapter 1) and the polarimeters attached to single channel photometers are usually designed to measure one or the other. Figure 7.14 shows schematically two arrangements of an analyser which can be used to

Fig. 7.14 Two, phase sensitive, single-channel analysers for the measurement of (a) the degree of circular polarisation or, (b) the degree of plane polarisation. The optical components are on the left with polarising action shown in the middle. The polarisation of the incoming light is shown at the top right followed by the changes in polarisation introduced by each component of the analyser. The output waveform of the detector is shown in the lower right as a function of the half wave plate position. (Published with permission from [385].)

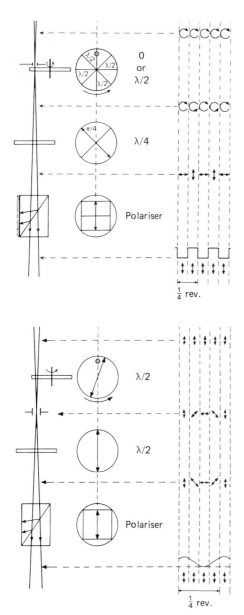

measure, phase sensitively, the degree of circular or plane polarisation [385]. The change in polarisation for either completely circular- or linear-polarised radiation is shown on the right with the phase of detection and modulation of the various retarders. The circular modulator is square-wave in its action while the linear modulator is sinusoidal. The final element is a polarising beam splitter which always delivers the same linear polarisation to the detector.

Using a rapid (100 Hz to 1 MHz) phase sensitive analyser preceding a single photomultiplier photometer operating in a photon counting mode, it is possible to achieve measuring accuracies < 0.01 per cent in the degree of polarisation since the results are insensitive to the slower changes of guiding and the atmosphere. The retarders are either plates of fixed dimensions which can be rotated, as in Figure 7.15, or a plate whose retardation can be varied piezo-optically or photoelastically [358, 357].

Analogues of these techniques exist at radio frequencies [387].

Gamma-ray and X-ray detectors

The detectors of such high energy and low flux photons as cosmic gamma- and X-rays operate in a strictly photon counting mode. The detection of gamma-ray photons through electron–positron pair production has already been described in Chapter 3 in connection with the Cos-B satellite.

At X-ray wavelengths the single channel analogue of the photomultiplier is the proportional counter of which an example is shown in Figure 7.15 [145]. The anode is a wire of about 0.01 mm diameter at a potential of several thousand volts. It threads a coaxial chamber filled with a mixture of gases. X-ray photons ionise the gas molecules, the electrons accelerate towards the anode and cause further ionisation which provides a 'gas multiplication' of up to some 10^5 electrons. The electron diffusion rate ($< 1\,\mu$s) is much more rapid than that of the positive ions (10–$100\,\mu$s). While only the electron pulse is actually measured the drift time of the ions establishes the detector dead time.

The composition of the filling gas is usually about 90 per cent noble gases and 10 per cent of an organic quench gas such as methane or carbon dioxide. A mixture of 90 per cent argon and 10 per cent methane is known as P-10. Apart from collisionally de-exciting the ions, the quench gas absorbs any UV photons generated by the ionisation thereby preventing the formation of electrons near the walls. The gas is normally at about atmospheric pressure. The lack of low level excitation states for the noble gases reduces the non-ionising collision losses and spurious counts from free electrons. The quench gap also helps to stabilise the counter against the ionising effects caused by small amounts of impurity gases [311].

The window of the counter is normally a thin sheet of a metal such as beryllium and/or an organic compound such as aluminised mylar. The responsive quantum efficiency is a function of the window transmission and the absorption of the gas. Figure 7.16 shows the quantum efficiency of an argon filled counter with a double layered window of $2\,\mu$m polycarbonate and $3.2\,\mu$m Teflon.

The number of electrons in the anode pulse is a function of the energy of the originally absorbed X-ray photon together with the statistical uncertainty introduced by the multiplication process and other effects in the counter. Empirically, the spread in anode energies, for a flux of monochromatic X-ray photons, is given by [81]

$$\text{FWHM} = 0.14/E^{0.5} \quad \text{keV} \tag{7.39}$$

where E is the monochromatic photon energy in keV.

Fig. 7.15 A proportional counter for X-rays. The rectangular tubing in the upper section acts as a collimator and supports a mylar window. The central wire in each section of the chamber is the anode. (Reproduced from [311].)

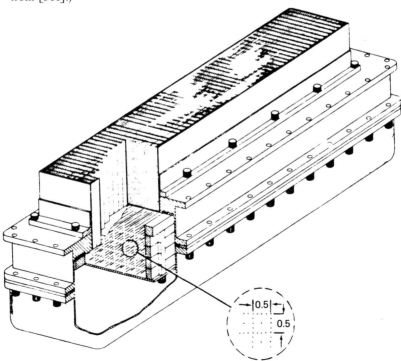

Infrared detectors

There are two basic types of incoherent detectors used for infrared observations [156, 259].

(i) *Thermal* detectors in which radiation from the source is absorbed on a black surface which increases the temperature of the sensor and thereby the phonon density which, in turn, alters the electrical characteristics.

(ii) *Photon* detectors in which absorbed photons cause electronic transitions which, again, alter the electrical properties. The electronic transitions are either *intrinsic* or *extrinsic* (impurity). In intrinsic excitation photons excite electrons from the valence band of a semiconductor to the conduction band to produce a free hole and an electron. In extrinsic excitation the incident photon ionises an impurity centre to create either a free electron and a bound hole or a free hole and a bound electron.

Photon detectors are of two types.

(1) *Photoconductive:* in which electrons are excited to the conduction band thereby changing its electrical conductivity.

(2) *Photovoltaic:* where photon generated electron hole pairs are separated at a p–n junction to give a detectable potential difference.

Table 7.1 lists the detectors most commonly used in astronomy together with the associated band-gap or excitation energies in electron-volts and the corresponding wavelength cutoff in sensitivity. The quantum efficiencies are approximate [156].

Fig. 7.16 Quantum efficiency of an argon filled proportional counter. The window of the counter has two layers, one consisting of 2 μm polycarbonate and the other of 3.2 μm Teflon. (Reproduced from [311].)

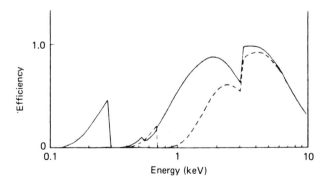

At infrared wavelengths the number of photons contributed by radiation from the telescope optics and the sky normally considerably exceeds the photon flux from the source. In consequence, the DQE is very small and the fluctuations in the background signal must be continually monitored. This makes simple analogue integration of the type discussed for optical wavelengths impractical.

Noise in semiconductors

The total mean square noise current in semiconductors, $\overline{i^2}$, has been found experimentally to have a form similar to that shown in Figure 7.17, where f is the frequency [237]. There are three distinct regions in the curve. Thermal noise exists equally at all frequencies and dominates at high frequencies. It is caused by the random motions of the charge carriers

Table 7.1 *Infrared detectors*

	Material	E (eV)	λ (μm)	q
Thermal				
	Ge:Ga	—	1–1000	0.9
Photoconductive				
	Ge:Hg	0.09	14	
	Si:As	0.049	24	0.5
	Ge:Cu	0.041	30	0.4
	Ge:Ga	0.011	120	0.5
Photovoltaic				
	InSb	0.23	5.5	0.7

Fig. 7.17 The frequency spectrum of noise current in semiconductors showing the three main contributions. (Reproduced from [237].)

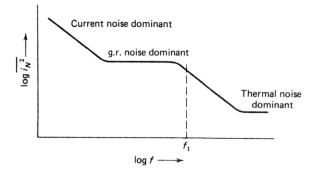

and the average power, P, is given by

$$P = kT\Delta f \quad \text{W} \tag{7.40}$$

where Δf is the band width in hertz, k is Boltzmann's constant and T is the semiconductor temperature. P is the noise power delivered to a circuit when its impedance is matched to that of the semiconductor. The open circuit noise would be $4P$.

In the middle frequency range the shot noise associated with the statistics of charge carrier recombination is important. This is known as g.r. noise (generation recombination) for short, and is discussed further below.

Current noise, usually called '$1/f$ noise' or flicker noise by analogy with electron tubes, is not well understood but appears to be associated with fluctuations in carrier density caused by surface leakage effects. Current noise becomes increasingly important at low frequencies and has the approximate form

$$\overline{i^2} = \text{constant} \cdot (I^2 \Delta f)/f \quad \text{A}^2 \tag{7.41}$$

where I is the average current flowing through the semiconductor.

It is these noise sources which ultimately limit the sensitivity of infrared detectors.

Noise equivalent power (NEP) [237]

A detector is normally characterised by its noise equivalent power (NEP). The NEP is defined as that power in the incident radiation which would produce an r.m.s. signal (current or voltage) from the detector equal to the r.m.s. noise signal (i.e. an s/n of unity). When the background radiation greatly exceeds that from the source the NEP is given as follows [156, 311]

$$NEP = (2hvB/q)^{0.5} \quad \text{W Hz}^{-0.5} \quad \text{photovoltaic, thermal} \tag{7.42}$$

$$NEP = 2(hvB/q)^{0.5} \quad \text{W Hz}^{-0.5} \quad \text{photoconductive} \tag{7.43}$$

where h is Planck's constant, v is the radiation frequency, q is the detector quantum efficiency, and B is the mean square power fluctuation in the background radiation absorbed by the detector. The higher NEP associated with photoconductive detectors comes from g.r. noise. While the carriers generated at a p–n junction are separated by the internal electric field, the carriers are not separated in the photoconductor and tend to recombine. The statistical fluctuation in the recombination has the same character and magnitude as photon noise.

Thermal detectors

The germanium bolometer operated at liquid helium temperature is the most widely used detector for infrared observations [105, 259]. It is sensitive from 1 μm to 10^3 μm. It consists of a gallium-doped germanium chip with a sensing area <0.1 mm^2. The sensing area is blackened for good absorption (>98 per cent) at wavelengths less than 60 μm. Metal leads are attached to either side of the chip by indium soldering techniques. In operation one lead is grounded electrically to a liquid helium cooled surface. The other lead is attached through sapphire to the same cold surface. The sapphire acts as a good electrical insulator but a good thermal conductor. The leads provide the only support for the detector and, consequently, the only path for heat conduction to the cold surface since the bolometer is operated in a vacuum. The *thermal conductance*, G (W K^{-1}), is adjusted by careful control of the lead dimensions.

The variation of the bolometer electrical resistance, $R(T)$, with temperature is given approximately by

$$R(T) = R_0(T_0/T)^4 \quad \Omega \tag{7.44}$$

where R_0 is the bolometer resistance at the temperature, T_0, of the cold wall and the fourth power depends somewhat on the doping level and degree of compensation.

A bias current is passed through the bolometer in series with a large load resistor. The bias current is adjusted to give an ohmic heating, H, of the detector, such that

$$H = 0.1 T_0 G \quad W \tag{7.45}$$

The responsivity, S, is defined as the r.m.s. change in resistance for unit change in the r.m.s. incident radiant power, and is given by

$$S = -0.7(R_0/T_0 G)^{0.5} \quad A^{-1} \tag{7.46}$$

which implies that the higher the thermal conductance the smaller will be the signal for a given input power. By the same token the thermal response time, τ, is

$$\tau = C/G \quad s \tag{7.47}$$

where C is the heat capacity (J K^{-1}) of the bolometer chip and its leads. Johnson noise associated with R_0 and the statistical fluctuations in the number of phonons flowing through the thermal conductance set a lower limit to the NEP in the absence of background of

$$NEP_{min} = 4T_0(kG)^{0.5} \quad W \; Hz^{-0.5} \tag{7.48}$$

where k is Boltzmann's constant.

The presence of background noise increases the NEP. When the photon noise from the background is dominant then the NEP is given by equation (7.42). Further, if the heating by the background radiation is large then, according to equation (7.46), the responsivity decreases while the thermal time constant increases.

The thermal conductance, G, controls the responsivity, S, the time constant, τ, the dynamic range in the presence of a high background, and NEP_{min}. The response time, τ, for the best sensitivity is of the order of 20 ms. A still longer time constant is excluded by $1/f$ noise. Table 7.2 lists the characteristics of a sensitive germanium bolometer and the responsivity and time constants for three levels of background (published with permission from [259]).

The minimum dimensions of a few tenths of a millimetre for the germanium chip are set by the increasing contribution of the contacts and leads to the heat capacity, C, and an actual degradation of performance below this size.

Photoconductive detectors
The current flowing through a photoconductor due to an incident power B is

$$i = q(BeG')/h\nu \quad \text{A} \tag{7.49}$$

where e is the electronic charge and G' is the *photoconductive* gain and is given by

$$G' = V_0 \mu \tau / l^2 \tag{7.50}$$

where V_0 is the bias voltage, μ is the mobility of the excited carrier ($\mu = v/V_0$ where v is the average velocity of the electron between scatterings), τ is the carrier lifetime, and l is the distance between the electrodes.

Figure 7.18 shows the biasing circuit used with a photoconductor. $R(L)$ is normally much smaller than the detector resistance $R(D)$ so that G' is almost independent of the current through the detector. The expression for the photoconductor NEP in the absence of background is

$$NEP = (9.22 \times 10^{-18}/\lambda q G')(T(L)/R(L))^{0.5} \quad \text{W Hz}^{-0.5} \tag{7.51}$$

where $R(L)$ is in Ω, λ is in m, and $T(L)$ is the temperature of the load resistor in K. Values of $R(L)/T(L)$ of about $10^6 \ \Omega \ \text{K}^{-1}$ are adequate for ground based telescopes, while values closer to $10^{11} \ \Omega \ \text{K}^{-1}$ may be more appropriate for a cryogenically cooled telescope in space.

Table 7.2 *Measured performance of a Ge bolometer under low background conditions at 10 μm*

Area A	0.28×0.29 mm^2			
Thickness i	0.25 mm			
Temperature T	1.5 K			
Thermal conductance G	1.1×10^{-7} W K^{-1}			
Background power Q	$<1 \times 10^{-8}$	2.7×10^{-8}	5×10^{-8}	W
Resistance R_0	7×10^6	3×10^6	1.7×10^6	Ω
Responsivity S_{max}	4.3×10^6	3×10^6	1.8×10^6	V W^{-1}
Time constant τ	8×10^{-3}	9×10^{-3}	11×10^{-3}	s
Modulation frequency f_0	20	18	15	Hz
Noise E_n	30	40	50	nV Hz$^{-1/2}$
NEP	7×10^{-15}	1.3×10^{-14}	2.5×10^{-14}	W Hz$^{-1/2}$
Photons s^{-1} at 10 μm	$<1 \times 10^{12}$	2×10^{12}	4×10^{12}	

Photovoltaic detectors

The InSb photovoltaic detector has almost completely replaced the PbS photoconductor and bolometer for astronomical observations at $\lambda < 5.5\,\mu m$ [156]. It is one of a group of semiconductors known as intermetallics. The width of the band gap at 77 K is 0.23 eV (i.e. a cutoff at 5.5 μm) and the electron mobility is about $30\,m^2\ V^{-1}\ s^{-1}$. The detector can be operated either in an unbiased photovoltaic mode or as a reverse biased photodiode. The detector resistance, $R(D)$, is about 1 kΩ and q is of the order of 0.7. The expression for the zero background NEP is almost identical to that for a photoconductive detector with $G' = 1$,

$$NEP = (9.22 \times 10^{-18}/\lambda q)(T(1/R(L) + 1/R(D)))^{0.5} \quad W\ Hz^{-0.5}$$

(7.52)

where resistances are in Ω and λ in m. By lowering the temperature of the detector and/or subjecting it to an infrared flash the value of $R(D)$ can be significantly increased thereby decreasing the NEP [363].

Background radiation

The background power reaching the detector is a function of the area–solid-angle product, $A\Phi$, of the telescope–detector combination, and the temperature and emissivity of the telescope optics and the atmosphere. $A\Phi$ is defined as

$$A\Phi = (\text{area of telescope primary})(FOV)$$

(7.53)

where FOV is the field of view of the detector or, more simply, the solid angle of the photosensitive area of the bolometer projected by the telescope optics onto the sky. For a point source the minimum angular projection of the detector is the diameter of the Airy disc, $2.5\lambda/D$. Hence, in this case,

$$(A\Phi)_{min} = 3.9\lambda^2 \quad cm^2\ sr$$

(7.54)

Fig. 7.18 Photoconductor biasing circuit.

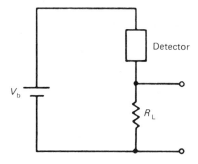

provided the diameter of the Airy disc exceeds the diameter of the seeing disc (as discussed in Chapter 4).

The precautions necessary to restrict the background emission from the telescope and the sky have been discussed in Chapter 3 in connection with telescope design. Figure 7.19 shows an infrared photometer in section and demonstrates how refrigeration is maximised by minimising the solid angle subtended by the telescope pupil [259]. To avoid excess radiation from the filters, they too must be refrigerated.

Fig. 7.19 Section of an infrared dewar where the angle subtended by the telescope pupil is minimised and the final optical elements are refrigerated in order to maximise refrigeration. (Published with permission from [259].)

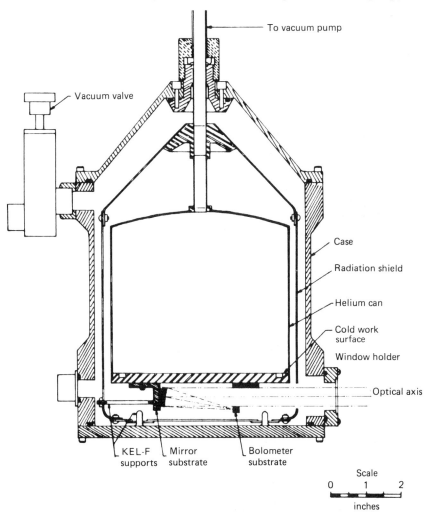

The background flux incident on the detector is limited by the bandpass of the filter or spectrograph which precedes the photometer. If $\Phi \ll 1$ steradian and the bandpass $\Delta\lambda \ll \lambda$, there is an approximate expression for the background power reaching the detector [156],

$$B = (\varepsilon F(\lambda, T) A\Phi\Delta\lambda \cdot t)/\pi \quad \text{W} \tag{7.55}$$

where $F(\lambda, T)$ is the Planck function (see Chapter 1) for an effective background temperature T, ε is the emissivity of the warm components of the optical system including the atmosphere, and t is the transmission of the cold optics. From equation (7.55) the NEP for the case of background limited performance can be estimated using equations (7.42) and (7.43),

$$NEP = 5 \times 10^{-12}\lambda((F(\lambda, T)\varepsilon t\Delta\lambda)/q\lambda)^{0.5} \quad \text{W Hz}^{-0.5} \tag{7.56}$$

for photovoltaic and thermal detectors, and

$$NEP = 7 \times 10^{-12}\lambda((F(\lambda, T)\varepsilon t\Delta\lambda)/q\lambda)^{0.5} \quad \text{W Hz}^{-0.5} \tag{7.57}$$

for photoconductive detectors. These expressions assume that $A\Phi$ is the $(A\Phi)_{min}$ of equation (7.54) and therefore must be considered minimum values since, normally, $A\Phi > (A\Phi)_{min}$.

Figure 7.20 shows examples of background limited NEPs based on equations (7.42) and (7.43) [156]. The calculations include an appropriate wave noise component as discussed in Chapter 2. In curve (a) the values $T = 300$ K and $\varepsilon t\Delta\lambda/\lambda q = 0.1$ are typical of broad band observations with a telescope at ambient temperature.

The emission from the telescope is ignored in curve (b) and the background corresponds to that expected from the interplanetary particles responsible for the zodiacal light [362]. Their characteristic temperature, $T = 300$ K, and their emissivity is a factor of 10^{-7} less than that of the Earth's atmosphere, i.e. $\varepsilon t\Delta\lambda/\lambda q = 10^{-8}$.

Curve (c) corresponds to the background expected from a telescope cryogenically cooled to 12 K with the same emissivity as a telescope at ambient temperature. The combination of curves (b) and (c) then corresponds to the NEP to be expected for a cryogenically cooled telescope in space. The background is dominated by emission from the zodiacal light for wavelengths shorter than 50 μm and by emission from the telescope at longer wavelengths.

In practice the emissivity, ε, can be measured directly. With an upward looking photometer the detector receives skylight from the primary which is seen imaged in the secondary. For a telescope designed specifically for low infrared background (see Chapter 3) the secondary has no surrounding support and is marginally undersized to prevent the detector from seeing the outer edge of the primary. A conical mirror at the centre of the secondary covers the image of the central hole in the primary and substitutes sky background in its place. As an infrared telescope should have no baffling the most serious source of telescope emission is the secondary mirror support spider.

Assuming that the detector has a linear response and gives a d.c. output power of P_{total} when the telescope points to the zenith, $P_{ambient}$, when the photometer is closed with a dark slide at ambient temperature, and P_{sky} when the photometer is removed from the telescope and pointed, independently, at the zenith then

$$\varepsilon = P_{total}/P_{ambient} \tag{7.58}$$

and

$$P_{telescope} = P_{total} - P_{sky} \quad \text{W} \tag{7.59}$$

Fig. 7.20 Detector noise equivalent power (NEP) for a range of detectors operating under different observing conditions. The solid lines are the best performance of bolometers and photodetectors for the given conditions and the pairs of dashed lines give the background limited performances for a photoconductor (upper) and photovoltaic detector (lower) under the three conditions (a), (b), and (c) discussed in the text. (Published with permission from [156].)

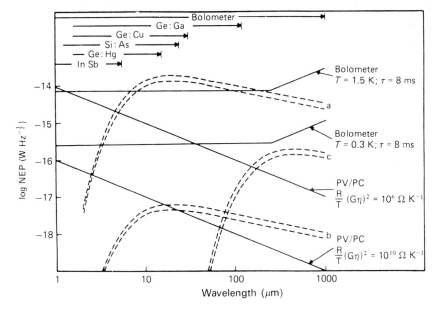

Signal-to-noise and DQE

An ideal infrared photometer should have two identical detectors fed by identical apertures one of which contains the source plus background while the other contains the background alone. The difference between the outputs of the two detectors would provide the source signal [157]. In practice only a single detector is used which is optically chopped between the two positions on the sky by an oscillating secondary (see Chapter 3). The chopping frequency is greater than the rate of variation in the background (other than photon shot noise). The signal-to-noise is given by

$$(s/n)_{out} = Sq/(Sq + 2NEP)^{0.5} \tag{7.60}$$

where S is the total source power incident on the detector during the observation in the case of a thermal detector, or the total number of photons in the case of a photon detector. The factor of 2 arises from the double measurement of the background compared with the source. The DQE is given by

$$DQE = q/(1 + 2NEP/Sq) \tag{7.61}$$

Coherent detection

Radio receivers [61]

With a dish antenna, radiation from the radio source of interest is delivered, in phase, to a feed horn at the focus (as discussed in Chapter 3). The feed horn and associated waveguide concentrate and duct the radiation to a receiver input. The latter is usually a stub in the waveguide whose length is tuned to match the input impedance of the receiver to that of the feed horn. The receiver converts the radio frequency power from one plane of polarisation into an electrical signal, normally a voltage, suitable for recording and monitoring with a meter.

Normally a receiver is only sensitive to a frequency band width, Δf, centred on the frequency of interest, f_0. The shape of the band width is usually Gaussian and Δf is defined as the difference between the frequencies at which the output voltage would equal the output voltage at f_0 if the input power to the receiver were halved. This is illustrated in Figure 7.21. If the receiver gain is linear Δf is measured between the frequencies giving output voltages which are 70.7 per cent of that at f_0.

Receiver noise is defined in terms of the receiver noise temperature $T(R)$ and it is the temperature at which a resistor would deliver the same noise power per hertz as the receiver into a matching circuit as defined in equation (7.40) [118]. A 1 K noise temperature is therefore equivalent to 1.38×10^{-23} W Hz^{-1} of available power [86].

A block diagram of a receiver and integrator is shown in Figure 7.22, while Figure 7.23 illustrates the processing sequence for a natural radio signal. The sections following the preamplifier and the mixer are the same for all frequencies while the preamplifier, mixer, and local oscillator are specific for each frequency.

Figure 7.23(a) shows the input signal with its spectrum in (b). Figures 7.23(c) and (d) show the components of the signal and the spectrum at f_0 to which the receiver is sensitive. Beat frequencies between 0 and Δf arise between the different frequency components contained within Δf. This beating introduces the low frequency modulation of wave noise shown in (c) and also sets an upper limit of $(\Delta f)^{-1}$ to the time scale for fluctuations in amplitude.

The output of the preamplifier is combined in the mixer with the more powerful signal from the local oscillator (LO). Energy is produced at the difference frequency which is often designed to be close to 30 MHz and passed on to the intermediate frequency (IF) amplifier. There are two frequencies, f_0 and f_2, which can produce the same beat frequencies

Fig. 7.21 Convention for the description of a receiver band width whose output voltage increases linearly with input voltage. (Published with permission from [61].)

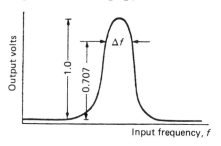

Fig. 7.22 Block diagram of a receiver and integrator. (Published with permission from [61].)

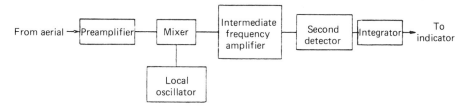

$(f_0 - f(\mathrm{LO}))$ and $(f(\mathrm{LO}) - f_2)$, where $f_2 = f_0 - 60$ MHz. f_2 is known as the image frequency and is normally suppressed by the selectivity or band width, Δf, of the preamplifier. When the selectivity of the preamplifier is insufficient then an image rejection filter is necessary.

The IF amplifier contributes the majority of the gain (typically 10^6) and its output is shown in (e) and has the same envelope as the wave form in (c) except that the mid-frequency is displaced to 30 MHz as shown in (f).

The second detector rectifies the negative going portions of the IF

Fig. 7.23 The changes in waveform and spectrum undergone by a signal with a continuous spectrum as it is processed by a noise-free receiver. The individual stages are discussed in the text. (Published with permission from [61].)

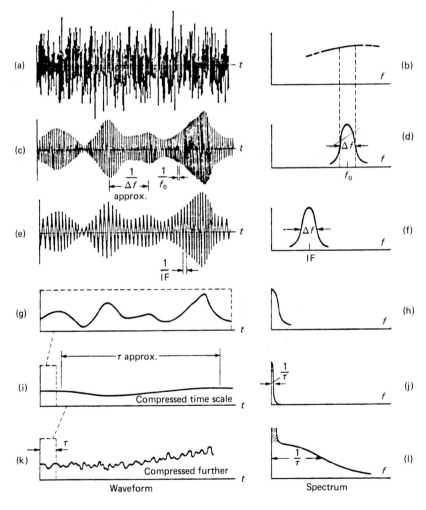

output and also filters out the higher frequency components to give the receiver output signal and spectrum shown in (g) and (h). Although the input power at f_0 was steady the output varies about a mean value at a rate which is no more rapid than the band width Δf. This is the wave noise discussed in Chapter 2 and in Chapter 6 in connection with the optical intensity interferometer.

Figures 7.23(i) to (l) illustrate the effects of the integrator time constant and long term drifts in the receiver gain and baseline. If the integrator has a time constant τ then, because of the nature of the wave noise or beating mentioned above, this will correspond to some $\tau \Delta f$ independent values of the receiver output. The r.m.s. uncertainty, σ, in the mean output signal is then given by

$$\sigma = (\text{mean signal})/(\tau \Delta f)^{0.5} \tag{7.62}$$

Wave noise is irreducible and independent of any additional noise injected by the receiver.

Receiver drift over periods $\gg (\Delta f)^{-1}$ contributes the low frequency peak on the expanded frequency scale shown in (l). The effect of drift is restricted by switching the receiver input between the telescope beam and a stable noise source such as a resistor at a known temperature (Dicke switch) [113], or, as in infrared observations, between two separate telescope beams (beam switching).

Infrared heterodyne

The technology of infrared heterodyne spectroscopy has developed principally around 10 μm, or 30 THz (tera $= 10^{12}$), because of the availability of stable CO_2 lasers and the high (90 per cent) transparency of the atmosphere at this frequency [338]. It is also fortunate that an astrophysically important molecule, ammonia, has its strong v_2 vibration–rotation band interleaved with the 25 to 33 THz laser bands of CO_2 and N_2O. Frequency coincidences of <3 GHz between laser lines and ammonia transitions are quite common. The same is also true for the v_7 band of ethylene centred near 28.5 THz. The molecule CO_2 itself is of course of interest in the study of planetary atmospheres [31].

Figure 7.24 shows a simplified schematic of a heterodyne spectrometer [32]. The telescope beam comes to a focus from the left and then passes through a NaCl beam splitter which transmits some 97 per cent of the radiation. The vertically polarised output from a CO_2/N_2O laser is expanded to match the focal ratio of the telescope and part is reflected towards the condensing lens. Only the central point of the laser beam is used when matching the telescope focal ratio in order to achieve as uniform an illumination as possible. The condensing lens before the photodiode operates at approximately $f/3$ giving a 60 μm diameter Airy disc which is imaged on an optimum point of the 150 μm diameter sensitive area of a HgCdTe photodiode.

The reverse biased HgCdTe photodiode acts as the photomixer and has an a.c. quantum efficiency of about 35 per cent at 10.6 μm [365]. The laser power incident on the photodiode is limited to some 0.5 to 0.8 mW which is sufficient to raise the laser induced shot noise in the photodiode to several times the preamplifier noise but still low enough to avoid reducing the effective quantum efficiency through saturation of the detector.

With the liquid nitrogen cooled GaAs-FET amplifier mounted close to the diode, quantum-noise limited detection can be achieved over a band width, Δf, of 1.5 GHz (10^9) [417]. The laser, which oscillates at a fixed frequency, can be calibrated to better than one part in 10^9. Variations in the radial velocity of the source caused by the rotation and orbital motion of the Earth can introduce frequency shifts of up to 10 MHz h^{-1} depending on the direction of the star. These shifts are compensated at an intermediate frequency prior to the spectrometer section represented by the filter bank.

The amplified IF output passes through a second stage of frequency conversion and into a bank of sixty-four 20 MHz filters. The filter outputs are integrated on analogue integrators which, in turn are sampled by a

Fig. 7.24 Components of an infrared heterodyne spectrometer which uses a CO_2 laser as first local oscillator. (Published with permission from [32].)

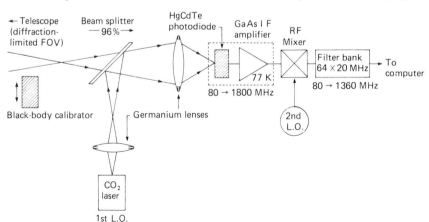

computer. The telescope beam is chopped between sky and sky plus source and the chopped signals from the filter bank are synchronously demodulated before integration.

The efficiency of the instrument is, in part, a function of the transmission of the optics and loss of source radiation through sky chopping. Only radiation polarised parallel to that of the laser is detected and only some 35 per cent of the source radiation incident on the diode produces an effective IF signal current in the photomixer. The net system detection efficiency is about 5 per cent.

8

Multi-channelled incoherent detectors

Photographic emulsion plus microdensitometer [110]

The photographic process [15]

Photographic emulsions have been used to record astronomical images for over 100 years and they are still the only detector which can fully exploit the large focal planes of wide field telescopes. The literature on the photographic process is extensive [110]. This section is restricted to a simple outline of image formation and measurement.

When a photographic emulsion is exposed to light some photons are absorbed by the silver halide crystals (often called grains). An absorbed photon excites an electron into the conduction band and leaves a positive hole in the valence band. Both the electron and the hole move about within the crystal until they become trapped or react chemically. Normally the electron becomes trapped at a crystal defect or at a site created by chemical sensitisation. Any silver ion which diffuses towards the electron will be neutralised to form a silver atom. A single silver atom is not stable but it can act as a centre for further electron and silver ion migration. When three or four silver atoms have accumulated they form a stable silver 'speck'. The formation of the speck constitutes what is known as a latent image.

In the presence of a suitable reducing agent (the developer) silver halide grains are progressively reduced to silver. While all the crystals would eventually be reduced, the rate of reduction is very much higher for crystals with latent image specks. In complete development, all of the silver ions are reduced to metallic silver within a crystal with a latent image. Thus, during development, there is an internal amplification of the order of 10^9 (i.e. some 10^9 silver atoms are produced for the three or four absorbed photons which originally excited electrons to the conduction band). This means that the larger the grain the greater will be the internal gain or speed of the emulsion (where speed is defined below in terms of photographic density).

The early growth of the silver speck is critical. In order to form a sufficiently stable nucleus a second silver atom must form at the site before the first atom dissociates. Clearly this nucleation stage is less efficient for low photon fluxes than during the subsequent growth of the silver speck to a developable size. Further, the electron can take part in other reactions that do not produce silver atoms, and this represents a loss of efficiency. It may react, for example, with oxygen and moisture in the surrounding gelatin or with impurity ions in the crystal, or it may simply recombine with a hole. Any additional photons which are absorbed by the crystal after the formation of the speck make no further contribution to image formation.

The 'pulse height' or grain size distribution associated with any particular emulsion is approximately log-normal (i.e. the log of the projection area approximates a normal distribution). Despite the fact that in a uniformly exposed image the average areas and the numbers of grains may vary by as much as a factor of 100, the product of the number and the areas of the grains normally only varies by factors of 2 or 3 [110].

Sensitising dyes are used to extend the sensitivity of emulsions to wavelengths beyond the 500 nm limit (the silver halide minimum excitation energy to the conduction band). The dyes fluoresce in the blue when illuminated by longer wavelengths. The sensitivity of conventional emulsions is limited at short wavelengths to about 250 nm by strong UV absorption in the gelatin, but they are sensitive to both X-rays and gamma rays.

Microdensitometer/microphotometers [110, 248]

Astronomical emulsions tend to be coarse grained compared with those used in portrait photography and the number of potentially developable grains is approximately 10^6 cm^{-2}. Wide field plates can be up to 40 cm on a side which implies 1.6×10^9 developable grains. Although, in theory, it should be possible to count and register individual grain positions in astronomical photographs, in practice it has not proved possible except in images of low density. Rather, the collective optical effect of the grains is measured over small areas. If light of intensity I_0 is projected through the developed image then the density, D, is defined as

$$D = \log_{10} I_0/I(T) \tag{8.1}$$

where $I(T)$ is the measured, transmitted, intensity. This can also be expressed as

$$D = -\log_{10} T = \log_{10} O \tag{8.2}$$

where T and O are the transmittance and the opacity, respectively.

Since the actual measurement technique in fact establishes the photometric properties of the image rather than measuring the density directly, the instrument is more accurately called a microphotometer. The measured density depends on the illumination geometry, and the degree of coherence and wavelength, of the illuminating radiation. Silver grains scatter a significant amount of the incident light as well as absorbing it. Density is called *specular* if the illumination and collection angles are small, and *diffuse* if one of these angles is large, as is shown in Figure 8.1. The ratio of specular to diffuse density is always greater than unity and is known as the Callier coefficient.

Figure 8.2 shows, schematically, the optical arrangement of a single channelled microdensitometer. The light A_1 is focussed by the field lens onto the field stop A_2 of a microscope ocular which forms an image of the field stop B_1 onto the emulsion (sample). The pickup microscope projects an enlarged image of the emulsion onto the scanning aperture B_3. The pickup condenser projects a Fabry image, the exit pupil of the scanning microscope A_3, onto the photomultiplier. The scanning aperture is adjusted for the appropriate resolution and the output of the photomultiplier is either digitised or displayed in synchronism with the motion of the emulsion normal to the scanning slit.

In order to sample the full spectrum of spatial frequencies in the image (i.e. down to structures of the same order as the average grain size), the slit

Fig. 8.1 Definitions of diffuse and specular densities in terms of the light transmitted and scattered by the emulsion.

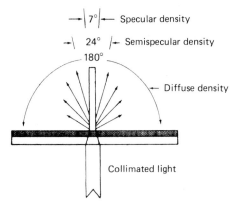

must be wide enough to provide incoherent illumination. This will be true if the following two conditions are met [110]:

(i) $s \geqslant 1 + \omega_{eml}/\omega_{lens}$ (8.3)

and

(ii) $W \geqslant 4\lambda/na_{coll}$ (8.4)

where $s = na_{coll}/na_{lens}$, W is the width of the field slit projected on the emulsion, ω_{eml} is the spatial frequency associated with the sampling interval on the emulsion and ω_{lens} is the maximum spatial frequency transmitted by the objective lens of the pick-up microscope which, if it is aberration free, is given by

$\omega_{lens} = 2na_{lens}/\lambda$ (8.5)

where na is the numerical aperture and λ is the wavelength of the analysing light. (The numerical aperture is the sine of the half-angle of the widest bundle of rays capable of entering the lens, multiplied by the refractive index of the medium containing the bundle, in this case air.)

In order to assure a uniform scale of density, most observatories have adopted a standard scale of diffuse densities (ANSI standard PH2.19-

Fig. 8.2 Optical system of a typical single beam microdensitometer. (Published with permission from [110].)

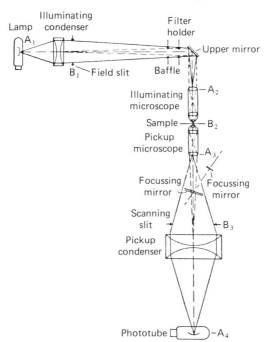

1959) based on a set of calibration plates [195]. As a result, plates measured with several microdensitometers with different illumination geometries should give the same diffuse densities.

The response of an emulsion to light is normally described by a plot of density against either log of the energetic incident flux or, preferably, the log of the exposure expressed in units of 1000 photons per square micrometre [202]. Historically a log–log rather than a linear plot is preferred because it mimics the log response of the eye. A typical characteristic curve is shown in Figure 8.3. It can be divided into three parts: a low light level 'toe', a linear portion, and a high light level turnover. In conventional photography the gradient of the curve is assigned the symbol γ, and it is a measure of the contrast in the image,

$$\gamma = dD/d(\log_{10} E) \tag{8.6}$$

where E is the exposure (J m^{-2}).

The gradient, γ', of the D vs. N plot is more useful in astronomy [273],

$$\gamma' = dD/dN \tag{8.7}$$

where N is the average number of photons per unit area corresponding to the density D. The characteristic curve is not unique, particularly in the region of the 'toe'. For an exposure consisting of a fixed number of incident photons, the density varies with exposure time, being weaker for the longer exposure even although the total number of photons is the

Fig. 8.3 A typical characteristic curve for a photographic plate indicating the effect of low light level reciprocity failure.

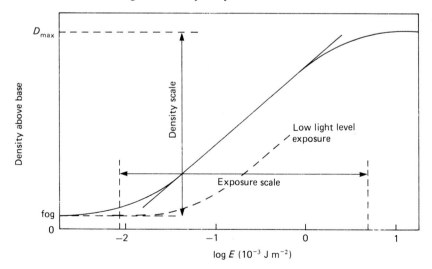

same. This effect is known as *reciprocity failure* because there is not a reciprocal relation between intensity and exposure when they are varied reciprocally. A low light level curve is shown dotted in Figure 8.3. (There is also a high flux reciprocity failure which is not so important in stellar astronomy.)

Both of the nonlinearities in the characteristic curve and reciprocity failure can be 'explained' in terms of the process of latent image formation outlined above. In general more than one silver atom must be produced in order to give a latent image. In consequence, at the low exposure levels implied in the 'toe' of the characteristic curve a larger proportion of the grains have only absorbed one photon than on the linear portion of the curve. When the arrival rate of the photons is of the same order as or smaller than the frequency with which single silver atoms disintegrate then the probability of developing a latent image is further reduced leading to low light level reciprocity failure. The overall quantum efficiency is also affected by the degree to which photoelectrons are lost through reactions other than those which produce silver atoms.

DQE of photographic emulsions [248, 273]

The amplitude of the output signal fluctuations from the microdensitometer is a function of the size of the sampling aperture, the density, and the average grain size of the emulsion. The granularity of the image is generally defined as the product of the root mean square density fluctuation, ΔD, and the square root of the sampling aperture area A. The parameter G, which is the square of the granularity, is then

$$G = A(\Delta D)^2 \quad \mathrm{m}^2 \tag{8.8}$$

It can be shown that [273]

$$\mathrm{DQE} = N\gamma'^2/G \tag{8.9}$$

where γ' and G correspond to the mean exposure of N photons per unit area. The result of more direct interest to the astronomer is

$$(s/n)_{\mathrm{out}} = N\gamma'/G^{0.5} \tag{8.10}$$

The granularity, $G^{0.5}$, contains not only a statistical component arising from the finite number of grains and the number of photons necessary to produce a latent image, but also the fluctuations in the photographic fog caused by thermal background and the development process. Normally, the zero point for density is defined with no plate in the beam. It is particularly important for calibration that the background be uniform across the plate. This is achieved by continually agitating the developer during development and by stopping development quickly and uniformly with a stop bath.

A measured DQE and $(s/n)_{out}$ curve is shown for an astronomical emulsion in Figure 8.4 [202]. The maximum in the DQE occurs almost at the start of the linear portion of the characteristic curve. This is to be expected since the number of grains without latent images decreases with increasing exposure. Since $(s/n)_{out} = N \cdot (DQE)^{0.5}$ the maximum in $(s/n)_{out}$ is broader and occurs at a larger value of N than for DQE. This implies that the best results are obtained with a fine grained (small $G^{0.5}$), high contrast (large γ'), emulsion.

Because of the nature of the latent image formation process DQE is a more useful quantity than responsive quantum efficiency for a photographic emulsion. Although one can define a quantum efficiency, q, which is the percentage of incident photons which produce silver atoms the actual number of grains developed also depends upon Q, the number of silver atoms required for a latent image speck. From Figure 8.4 the maximum values of DQE are about 2 per cent.

In contrast to the DQE and $(s/n)_{out}$ curves for the silicon diode and the photomultiplier in Figures 7.5 and 7.9 those shown in Figure 8.4 for photographic emulsions indicate that optimum exposure conditions occur for only a limited range of N (i.e. emulsions have a limited dynamic range). To this limitation can be added the other three already discussed,

Fig. 8.4 The characteristic curve, DQE, and $(s/n)_{out}$, for an untreated IIIa-J(1L5) emulsion developed in D-19 developer. (Published with permission from [202].)

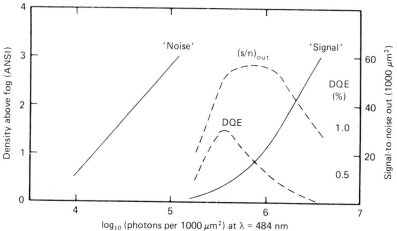

$\log_{10}$ (photons per 1000 μm^2) at $\lambda = 484$ nm

i.e. reciprocity failure, the threshold introduced by the toe of the characteristic curve, and low DQE. Each of these deficiencies can be remedied, but often only at the expense of accentuating one or more of the others.

Although a photomultiplier as a single channel detector has a much higher DQE than a photographic emulsion, it loses the advantage rapidly when a large area of image or spectrum has to be observed. Sometimes known as simultaneity gain or multiplex advantage [284], the advantage of the multi-channel detector over the single channel detector can be expressed as

$$\text{Simultaneity gain} = nQ/Q' \tag{8.11}$$

where Q and Q' are the DQEs of the multi-channelled and single channelled devices, respectively, and n is the number of independent channels in the multi-channelled device. Equation (8.11) ignores detailed advantages enjoyed by the single channel detector such as superior dynamic range, and more precise calibration.

Hypersensitising [15, 195, 293]

This treatment is directed towards the reduction of low light level reciprocity failure (LIRF), and to an increase in the DQE. Normally two classes of exposure are recognised. In Class I the signal-to-noise ratio in the image is sufficiently high that the toe or threshold of the characteristic curve affects the exposure time more critically than reciprocity failure. In Class II exposures the light level is lower and reciprocity failure is more serious.

The most widely used hypersensitising technique for Class I is a pre-exposure flash. A short, uniform, low level flash will produce single silver atoms which become developable with the absorption of photons from the source. The flash has the effect of moving the zero point of the exposure past the toe of the characteristic curve as is shown schematically in Figure 8.5. Gains of up to a factor of 20 have been achieved for some emulsions in this way.

Reciprocity failure is due largely to chemical reactions which cause single silver atoms to disintegrate or free electrons to be lost before the generation of a second atom by an absorbed photon. Hypersensitising in the case of Class II involves the removal of oxygen and residual moisture by one or more of the following methods, viz., evacuation, baking, or soaking in an inert or reducing atmosphere such as hydrogen [354].

Cooling of the emulsion during exposure is also effective but it is difficult to achieve at the telescope without condensation.

Table 8.1 summarises the effective hypersensitising techniques for a variety of emulsions used in astronomy (published with permission from [195]). Pure hydrogen is considered dangerous under some circumstances. Forming gas which consists of about 92 per cent nitrogen and 8 per cent hydrogen is much less hazardous and almost as effective as pure hydrogen [354]. The spectral responses of some of the emulsions to an incandescent light filtered to mimic the average spectral energy distribution of sunlight are shown in Figure 8.6 [195]. Figure 8.7 shows the improvement in DQE of the 127-04 emulsion with hydrogen soaking [249]. The numbers indicate the number of hours of soaking and the results are compared with an untreated plate and with one which was evacuated. The loss of DQE after 2 hours of soaking is caused by an increase in background noise.

Photographic enhancement and unsharp masking [147, 148, 268, 269]

This treatment improves the DQE over a limited range of the characteristic curve for a restricted range of spatial frequencies. A photographic emulsion is a three-dimensional recording medium in which unsaturated images tend to form close to the exposed surface while silver grains from chemical fog, which is inevitable in the hypersensitising process, are randomly distributed. The effect is more pronounced for finer grained emulsions of the IIIa type and arises from the turbid nature of the silver halide suspension. The concentration of grains towards the surface

Fig. 8.5 The effect of a pre-exposure flash on the characteristic curve.

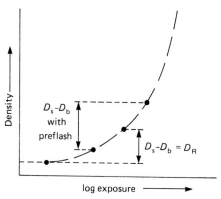

Table 8.1 *Summary of hypersensitisation treatments*

Type of Kodak spectroscopic plate	Gas treatments					Liquid treatments			Others	
	Baking (usually 55 to 75°C)			Soaking (20°C)						
	Air	N_2	Forming gas	N_2	H_2	H_2O	NH_4OH	$AgNO_3$	Evacuate	Cool
103a-O	e	E	E	E	E+	e			E	e
103a-D		E								
103a-E		E		e	E					
098	e	E		E	E	E	e		E	
IIa-O		E+	VE+	E	E+	e			E	
IIa-D		E		E	E				e	
IIa-F		E	VE+		VE				e	E
IIIa-J	e	E+	VE+	E	VE+	To	be	avoided	E	E
IIIa-F	e	E	VE+	E	VE+	To	be	avoided	E	
IV-N	e	e	E	e		E	E+	VE+	E	
I-N					E	E	E	VE+	E	
I-Z			VE+			E	E	VE+		E

Key to Table
VE   Very effective   e   Little effect
E   Effective   (blank) No result known

Table 8.1 cont. *Mechanisms of hypersensitisation*

Type of Kodak spectroscopic plate	Gas treatments					Liquid treatments			Others	
	Baking (usually 55 to 75°C)			Soaking (20°C)						
	Air	N_2	Forming gas	N_2	H_2	H_2O	NH_4OH	$AgNO_3$	Evacuate	Cool
Further chemical sensitisation	✓	✓	✓		✓					
Reduction sensitisation	✓	✓	✓		✓	?	?	✓		
Remove O_2		✓	✓	✓	✓				✓	
Remove H_2O	✓	✓	✓	✓	✓				✓	
Raise Ag^+ conc.						✓		✓		✓
Reduce recombination				✓						
Increase Ag^- stability				✓						

Key to Table
✓ Preferred by at least one user.

Fig. 8.6 The spectral response of various emulsions to an incandescent light filtered with a Corning 5900 glass filter to approximate the spectral energy distribution of the Sun. The UV limit is due to the source. Development and exposure conditions are indicated. The intensity ratios between successive steps is $\sqrt{2}$. (Published with permission from [195].)

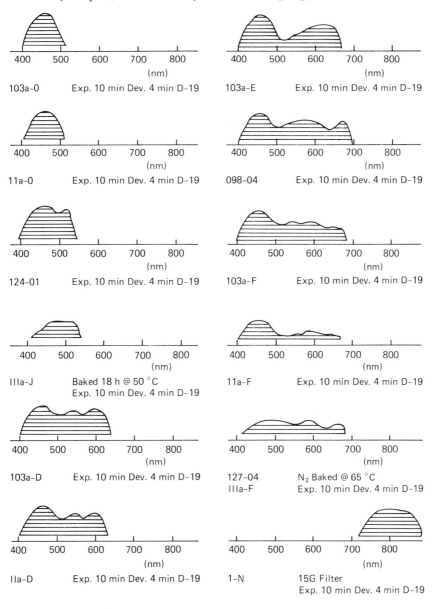

103a-0   Exp. 10 min Dev. 4 min D-19

11a-0   Exp. 10 min Dev. 4 min D-19

124-01   Exp. 10 min Dev. 4 min D-19

IIIa-J   Baked 18 h @ 50 °C
Exp. 10 min Dev. 4 min D-19

103a-D   Exp. 10 min Dev. 4 min D-19

IIa-D   Exp. 10 min Dev. 4 min D-19

103a-E   Exp. 10 min Dev. 4 min D-19

098-04   Exp. 10 min Dev. 4 min D-19

103a-F   Exp. 10 min Dev. 4 min D-19

11a-F   Exp. 10 min Dev. 4 min D-19

127-04
IIIa-F N_2 Baked @ 65 °C
Exp. 10 min Dev. 4 min D-19

1-N   15G Filter
Exp. 10 min Dev. 4 min D-19

also depends on the wavelength of the radiation to which the emulsion was exposed. Red light tends to penetrate further than blue into the emulsion.

A contact print made from the original plate using a diffuse light source for illumination and a high contrast film for the copy enhances the images in the near surface layer while suppressing the noise contribution from the chemical fog grains.

Figure 8.8 shows, schematically, sections of two emulsions with identical exposures and development with the fog (asterisks) and image (circles) grains identified [269]. The left-hand emulsion is illuminated by a distant point source and gives equal weight to each grain in the contact print. The emulsion on the right is illuminated diffusely. Grains close to the substrate cast wide shadows, which tends to contribute a structureless neutral density, while surface grains are recorded, considerably enlarged,

Fig. 8.7 Variation of DQE with number of hours of hydrogen soaking is shown for a 127-04 emulsion. (Published with permission from [249].)

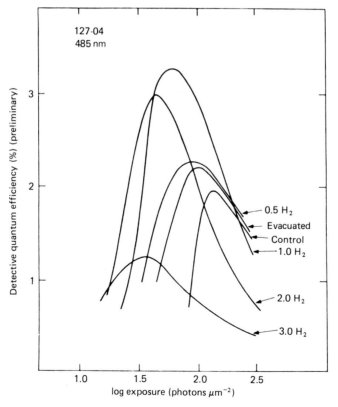

on the copy. The procedure increases contrast against the sky background for a limited range of the characteristic curve at the expense of enhanced granularity. The technique is particularly effective in improving the visibility of low surface brightness features.

Unsharp masking is a two-dimensional filtering technique which enhances the information at high spatial frequencies both in the core of a dense image as well as in its fainter outer regions and records it in a single print. This is achieved by suppression of lower frequency components in the image. In the analogue technique a negative mask of the original plate is made using diffuse illumination and a finite separation between the original emulsion and that of the print [147].

Photographic emulsion plus image intensifier

A range of image intensifiers is now available which covers wavelengths from X-rays to the infrared [410]. In combination with a photographic emulsion it is possible to benefit from the high quantum efficiency and the wider spectral sensitivity of photocathodes. It overcomes the light threshold at the toe of the characteristic curve because a single photon absorbed at the photocathode produces a scintillation of

Fig. 8.8 The mechanism of image amplification through diffuse illumination. The solid points are image grains and tend to be formed close to the exposed surface while fog grains shown as asterisks form throughout the emulsion. Parallel illumination on the left weights each grain equally while the diffuse illumination in a contact print on the right enlarges the image grains but leaves the deeper, fog grains undetected. (Published with permission from [269].)

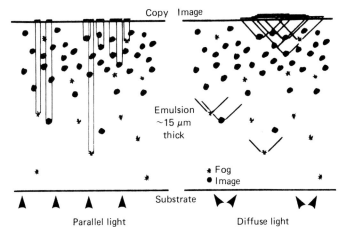

many photons at the phosphor. By shortening exposures it also limits the effect of reciprocity failure. These gains are made at the expense of dynamic range and image distortion [314], and they are limited by the area of available photocathodes.

There are several basic types of image intensifier available and these are illustrated in Figures 8.9 and 8.10 [245]. A photoelectron released from the photocathode by an incident photon is accelerated either directly to a phosphor or first to a microchannel dynode where it is multiplied before striking the phosphor. The phosphor has a thin aluminium coating on the side towards the photocathode to prevent light from the phosphor scintillation being seen by the photocathode. The electronic image is focussed onto the phosphor in three possible ways [84]:

    (i) by an axial magnetic field,

    (ii) by suitably shaped electrostatic potentials, or

    (iii) by proximity focussing over a small distance.

The phosphor scintillations are imaged onto the emulsion either through a transfer lens or through a fibre optic bundle in direct contact with both the phosphor and the emulsion. The overall gain can be improved by fibre-optically coupling several intensifiers to form a multistage device.

Radial and longitudinal dispersions in the velocities of the photoelectrons emerging from the photocathode introduce a spread or defocussing, Δ, at the phosphor. For a proximity focussed device Δ is given approximately by

$$\Delta = 2L(V_r/V)^{0.5} \quad \text{m} \tag{8.12a}$$

and for a magnetically focussed tube by

$$\Delta = LV_1/V \quad \text{m} \tag{8.12b}$$

where V_r and V_1 correspond to the radial and longitudinal components of energy of the emerging photoelectron. V is the energy in electron-volts gained by the photoelectron by acceleration in the potential across the gap L. Typical values of V_r and V_1 are 2 eV and the values increase as λ^{-1} with photon energy. For a magnetically focussed tube the length L giving a good focus is

$$L = 1.054 \times 10^{-5}(V^{0.5})/B \quad \text{m} \tag{8.13}$$

where B is the magnetic flux density in teslas.

The gain of an image intensifier is normally quoted as a luminous gain which is equal to the ratio of luminous flux emerging from the phosphor to the flux incident on the photocathode. Such a definition depends on the

Fig. 8.9 Schematics of the two main types of focussed image intensifiers: (a) electronic focussing, (b) magnetic focussing. In both cases photoelectrons are accelerated by the high voltage applied between anode and cathode. (Reproduced from [78].)

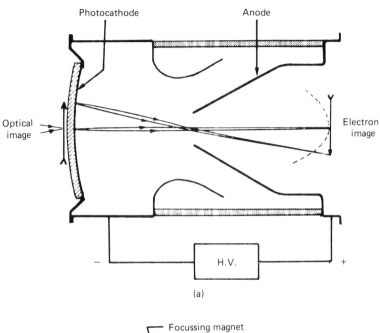

(a)

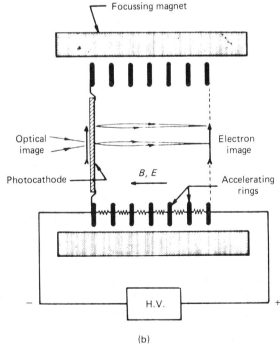

(b)

quantum efficiency of the photocathode material as well as the internal gain of the intensifier. It is more useful from the astronomer's point of view to separate these two. The photon yield is the average number of photons emitted by a phosphor scintillation for each photoelectron from the photocathode. An efficient phosphor emits some 0.04 photons eV^{-1}. The other phosphor quantities of importance are its time decay characteristics or persistence and its colour.

Each type of image intensifier has its own specific advantages and disadvantages [314].

Magnetically focussed intensifiers

These have a flat photocathode and tend to produce better quality images and have a higher quantum efficiency than the others. Unfortunately the magnetic yolk can be both bulky and heavy. Although the field can be provided by a permanent magnet the best images result from an electromagnet or solenoid. The latter consumes considerable power and the intensifier must be cooled in order to maintain a low dark current. The magnet must extend beyond the photocathode in order to provide a uniform field which is awkward for fast spectrograph optics. Magnetically focussed intensifiers suffer from some degree of S-distortion which means that a straight line such as a spectrum develops a shallow S shape which is a function of the accelerating voltage.

For a large enough magnetic field the radius of gyration of the electron can be made smaller than, say, the emulsion or optical resolution, which gives an 'infinite depth of focus'. This allows the freedom to choose the most convenient figure for the photocathode or emulsion surface. The radius of gyration, R, is given by

$$R = 2.692 \times 10^{-6} (V_r^{0.5})/B \quad \text{m} \tag{8.14}$$

where V_r is the transverse component of the photoelectron energy in eV, and B is the magnetic flux density in teslas. A typical value for V_r is 2 eV [78].

Electrostatically focussed intensifiers

These generally have strongly curved photocathodes and phosphors to match the 'pinhole' electron optics. The tubes are normally supplied with fibre optic face and rear plates to match them to a flat field. The glass in the fibre optics absorbs wavelengths < 400 nm; further, the combination of the front and rear plate fibre optics and the electron optics produces a marked variation in radial sensitivity across the photocathode. The sensitivity is usually about twice as high at the centre as it is at the edge.

Proximity focussed intensifiers

These are extremely compact because of the small distance over which the electrons are accelerated (see for example the UV converter in Figure 8.14). These intensifiers normally have flat photocathodes but, in theory, they can be curved to match a focal plane. The only restriction on the size of the photocathode is the ability of the front and rear plates to withstand atmospheric pressure. A simple proximity focussed tube with no dynode multiplication, known as a diode, has a luminous gain of about 20 to 50. Such devices often provide the first, UV sensitive, stage in a detector. Diodes tend to have a high stray light level which is contributed by light passing through the semitransparent photocathode and being reflected from the aluminium backing of the phosphor [206].

Coupled to a *microchannel plate (MCP)* electron multiplier, the proximity focussed intensifier can give photon gains of up to 10^6. MCPs are porous glass wafers some 2 to 4 mm thick with internal pore diameters of 12 or 25 μm on 15 or 32 μm centres. Sectional schematics of two MCPs are shown in Figure 8.10 [245]. The resistance of the glass is high enough to sustain a potential of about 1500 V between the faces [245, 382]. Photoelectrons are produced either from a semitransparent photocathode in the case of visible radiation or from an opaque photocathode deposited directly onto the surface of the MCP in the case of the vacuum UV or supported on a mesh. The photoelectron enters one of the pores or channels of the MCP where secondary electrons are generated at each collision with the wall.

The multiplication saturates when the space charge in the electron pulse inhibits the generation of further secondaries. This is also known as the avalanche mode and leads to a sharper pulse height distribution than from, say, a photomultiplier, but the resistance of the pore walls introduces a recovery or dead time of several milliseconds for the channel while the charge lost in the pulse is being replaced. The pores in modern MCPs are either curved or chevron shaped. This prevents direct feedback of positive ions from the anode side to the photocathode which helps to prolong the life of the photocathode and reduce the ion scintillation background.

The electrons emerging from the multiplier are proximity focussed to the phosphor. This latter separation has to be kept to a minimum because mutual repulsion causes the electron cloud to diverge with a cone angle of about 10°. In consequence the FWHM of the phosphor scintillations from a MCP tends to be larger than for either magnetically or electrostatically focussed multi-stage tubes operating at the same gain. The ratio of the

Fig. 8.10 Two image intensifiers with microchannel plates for electron multiplication. The upper one is proximity focussed while the lower one has an electrostatically focussed section.

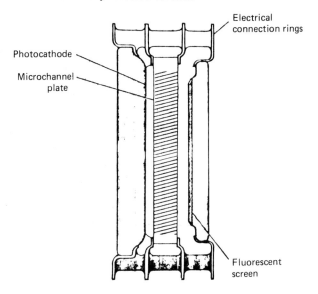

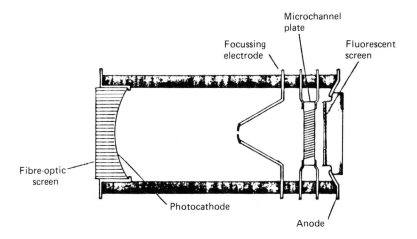

areas of the pore openings to the total pore area is about 0.5 unless the pore inputs have been funnelled to increase their area. Although secondary electrons produced in the inter-pore region are likely to be drawn into one of the nearer pores there is probably a variation of DQE on the scale of the thickness of the pore walls which would produce aliasing when the image contains structure of a similar spatial frequency.

The best quantum efficiency quoted for MCPs is usually less than 15 per cent, which is poorer than that of magnetically focussed tubes. This is due in part to the low sensitivity of the inter-pore region and the need to avoid ion feedback. For many applications involving signal generating detectors (discussed later in this chapter) it is important to preserve good pulse height characteristics in the charge pulses associated with photon events. When photoelectrons strike the interpore region of the MCP face those secondary electrons which do fall into a pore produce smaller charge pulses than those generated directly in the pore. This has the effect of broadening the pulse height distribution towards lower energies which makes photon events harder to detect.

When the photosensitive material is deposited directly onto the MCP face the primary electrons are normally insufficiently energetic to reach the pores from the interpore region. An aluminium ion barrier of up to 7 nm thickness is frequently deposited on the MCP face. This has the effect of limiting ion feedback to the photocathode, helps to prevent caesium entering the pores during preparation of the photocathode, and inhibits secondary electron emission from the interpore region. All of these precautions lower the quantum efficiency.

Coupling of phosphor output and detector

With a nominal $f/2$ coupling lens (i.e. operating at $f/4$) only 1 per cent of the light emerging isotropically from a phosphor scintillation would reach the plate, whereas for fibre optic coupling the figure is more like 50 per cent. None the less, the former is often preferred, particularly with magnetically focussed tubes. The photocathode is often at ground potential to avoid corona to the surroundings. In consequence, the phosphor is at a high positive potential which means that a large potential would build up across the fibre optic [107]. The image quality with lens coupling is generally better than with fibre optics. In particular the fibre optic boule is formed from a series of hexagonal bundles and shear at the edges of the bundles often leads to image discontinuities as well as a 'chicken wire' effect in the image caused by inefficient light transfer at the bundle edges. Lens coupling has the practical advantage of making plate changes simpler.

If the coupling efficiency between phosphor and plate is ε, the photon gain of the image intensifier is Γ, the area of the scintillation image on the plate is a, and the average grain area is α, then, in order to have a high probability of producing at least one developable grain per photoelectron we require

$$1 = (\varepsilon\Gamma\alpha/a)\text{DQE} \qquad (8.15)$$

Taking $\text{DQE} = 0.01$, $\varepsilon = 0.01$, and $\alpha/a = 0.1$, then the minimum value of Γ is

$$\Gamma_{min} = 10^5 \qquad (8.16)$$

This will be true only if the colour of the phosphor matches the peak of the emulsion spectral response. The spectral distribution of emission from the P-11 and P-20 phosphors which are those most commonly used are shown in Figure 8.11.

In spectroscopy where the light levels are low it is important that $\Gamma > \Gamma_{min}$. In direct photography where the sky background provides a uniform low light level illumination and where it is important to maintain a wide dynamic range it is better to have $\Gamma < \Gamma_{min}$. The photon gain/stage of electromagnetic and electrostatic image tubes is of the order of 10^2.

Fig. 8.11 Typical absolute spectral response characteristics of aluminised phosphor screens.

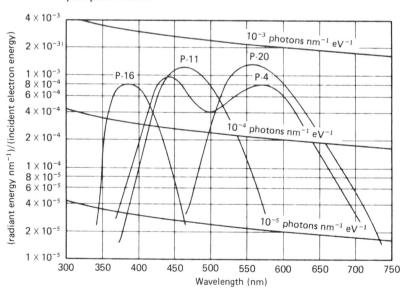

Consequently, a minimum of three stages is generally best for spectroscopy, or a single stage for a MCP intensifier.

Clearly, any advantage of the image tube plus photographic plate compared with the photographic plate alone is lost if $(\Gamma q \varepsilon) < \text{DQE}$, where the values of q, the quantum efficiency of the intensifier photocathode, and the DQE of the emulsion are for the wavelength of interest.

Electronography [76, 236]

When a nuclear emulsion is substituted for the phosphor in an image intensifier an electronic camera is the result. Such a device enjoys the advantage of high photocathode sensitivity with the wide dynamic range and the fine grain of nuclear emulsions without having the coupling losses and undesirable phosphor characteristics of the intensifier and photographic plate combination. Each photoelectron creates a trail of some ten grains in the emulsion. The emulsion response is highly linear with no reciprocity failure [192].

The principal disadvantages of electronography are the difficulties of introducing the emulsion into the envelope of the intensifier without contamination of the photocathode, the measurement and calibration of plate densities, and the uneven quality of available nuclear emulsion plates.

There are two principal types of electronic camera [78, 165, 283, 421]. In the first, the emulsion is introduced directly into the intensifier envelope and, in the second the emulsion is separated from the intensifier section by a membrane which is permeable to the accelerated photoelectrons. One example of the first type of camera and two of the second type are shown in Figure 8.12 together with a combination Schmidt/electronographic camera. The latter was operated as part of a spectrograph on the lunar surface (21–3 April, 1972) during the Apollo 16 mission. It obtained far UV objective spectra and direct images of (20° diameter) selected regions over some 8 per cent of the sky which included the Large Magellanic Cloud (LMC) [77].

Both the photocathode and the emulsion can be replaced in the case of the caméra électronique in Figure 8.12(a). The photocathode is isolated from the emulsion by the coin valve until the chamber containing the emulsion has been evacuated and the emulsion has been refrigerated long enough to avoid further significant outgassing.

In the case of the two cameras with membranes, one uses a 4 μm thick mica window while the other uses a 7.5 μm Kapton foil. The mechanics for

Fig. 8.12 Four electronic camera designs, in three of which the photocathode is protected from outgassing of the emulsion either by refrigerating the emulsion or by a thin membrane. The fourth camera for use in the ultraviolet has an opaque photocathode. (a, c, and d reproduced from [78, 236, 283] and b with permission from [165].)

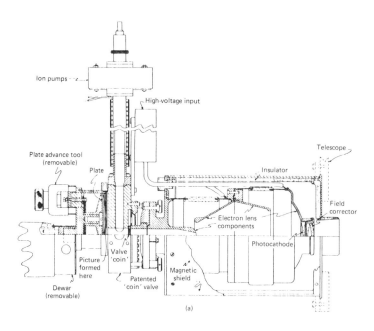

(a)

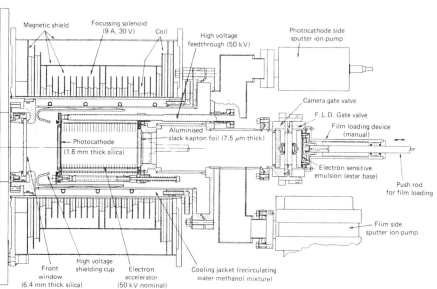

(b)

the introduction of the film based nuclear emulsion is shown for both. The membranes are protected from rupture by atmospheric pressure by maintaining a good vacuum on the film side at all times. The photoelectrons are accelerated to energies between 40 keV and 50 keV, with about 25 per cent of them being lost in the mica. The mica also introduces a spread of about 10 keV in energy as well as some scattering. The spread in energy introduces an associated spread in the

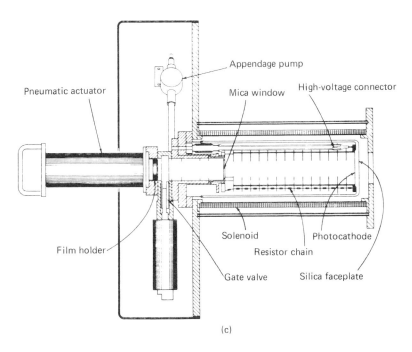

(c)

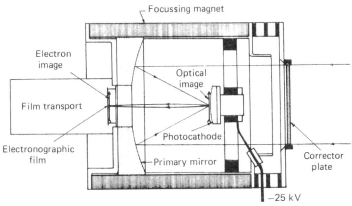

(d)

number of grains produced in the emulsion for each incoming electron, which has the effect of reducing the DQE compared with devices, such as the caméra électronique, in which the emulsion is exposed directly to the electrons.

Signal generating detectors [383]

Multi-channelled, signal generating detectors provide an output voltage level which is a function of the incident light. Each channel is read out in turn through a common amplifier and the image is reconstructed knowing the order in which the channels were read out.

Analogue and photon counting television cameras [261]

The operation of a television camera is shown schematically in Figure 8.13. The basic components are a photocathode or photosensitive surface from which incident radiation releases photoelectrons or charge carriers. In the case of photoelectrons these are accelerated to a thin target where they release secondary electrons thereby leaving a positive charge pattern in the target.

Fig. 8.13 The principal components of a television camera including details of the reading-beam/target interaction.

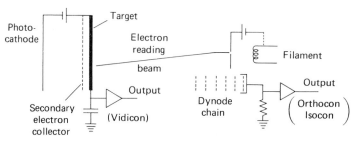

TV camera schematic

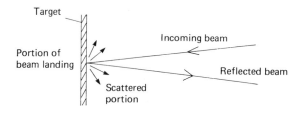

Reading-beam/target interaction detail

Normally the target is scanned from the rear in a raster by the electrons from an electron gun. The electrons evaporate from a hot filament and are accelerated to a few volts by a cylindrical anode. The beam is defined by a pinhole and continues towards the target. The raster scan is generated by deflecting the beam with varying magnetic fields or electrostatic potentials. The electrons in the beam arrive at the target with very low energy. Some of the beam lands and neutralises the positive charge while simultaneously altering the target potential, some of the beam is scattered by the positive potential well, while the majority of the beam is reflected and returns towards the anode.

A signal can be derived in three ways.

  (i) Orthocon: the reflected beam is intercepted, and amplified, in a dynode chain. This signal is not very suitable for low light level observations because it produces a negative image (i.e. the highest signal current corresponds to zero positive charge in the target) in which the full noise of the reading beam appears in the dark parts of the image.

  (ii) Isocon: the scattered electrons are separated from the reflected beam, and amplified in a dynode chain. While such devices are sensitive and have been used successfully in astronomy, they are bulky and, for the available cameras, the finite resistance of the thin (10 μm) glass target causes the image to spread during long exposures.

  (iii) Vidicon: this is the most compact and widely used camera in astronomy and has been particularly successful as an analogue device on satellite observatories. The signal is derived from the change in potential of the target during the raster scan. It is possible to take long exposures when the target is a good insulator. Unlike the Orthocon and Isocon, the Vidicon has no internal gain, hence care must be taken in the design of the video amplifier.

The SEC vidicon, a 2D analogue integrating detector

Figure 8.14 shows, in section one of the four cameras which operate in an analogue integration mode on the International Ultraviolet Explorer (IUE) Satellite launched in 1978 [100]. They obtained some 25 000 spectra in the first three years and are still in operation. The first element is a 40 mm diameter UV sensitive diode converter with a caesium telluride photocathode behind a magnesium fluoride window. The converter is followed by a secondary electron conducting (SEC) vidicon which has a $10\,\mu m$ thick potassium chloride target from which the secondary electrons are swept to a 50 nm thick aluminium signal plate by a 12 V bias. This leaves a positive charge in the insulator. As with all television cameras operating in an integrating mode the reading beam is turned off during an exposure to allow the charge pattern to build up in the target. During readout the reading beam moves in steps rather than continuously across the target.

Fig. 8.14 Detail of one of the four SEC integrating vidicon television cameras used on the International Ultraviolet Explorer satellite. (Reproduced from [100].)

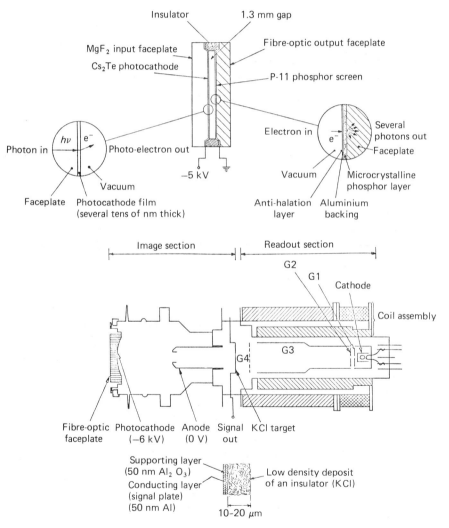

SEC vidicons have seen extensive use in integrating television detecting systems for both space and ground based astronomy [163, 305]. The highly insulating target means that charge can be integrated for many hours without degradation due to leakage. The potassium chloride has poor mechanical strength and can perforate and blow off if too highly charged. This can introduce complications in direct photography when a field contains one or more bright stars.

Photon counting television systems [42, 50]

Figure 8.15 shows, in section, the principal components of a ground based photon counting system which incorporates a television camera [42, 43]. The camera, a plumbicon, is lens coupled to the phosphor output of a four-stage magnetically focussed image intensifier. The plumbicon was chosen because it has low lag (see below), good resolution and a spectral response which matches the P-11 phosphor output. The photon gain is high enough that individual scintillations tend to give saturated target images which are sensed as pulses of approximately equal amplitude on the video output. The video pulse appears on more than one scan line and it is possible to find the centre of a scintillation to better than one scan line using a digital centring logic in the output circuit. This allows some of the resolution lost in the intensification and readout processes to be recovered. Ion events can also be recognised and eliminated because of their greater size.

The technique is illustrated in Figure 8.16 which shows, schematically, the way in which events are sensed above a threshold and centred while, at

Fig. 8.15 Section of the intensified television camera detector used in the Boksenberg photon counting system. (Reproduced from [42].)

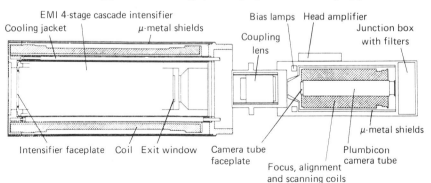

the same time, sensing and eliminating the larger ion events. The maximum readout rate is the standard TV scan rate of 30 Hz which means that pile-up errors become important at quite low photon arrival rates. This limit is further restricted by the 'dead' space associated with the area of a scintillation. The detectors in the Faint Object Camera (FOC), of the Hubble Space Telescope, will be photon counting systems using a three-stage image intensifier lens coupled to a SIT (EBS) television camera (see below) [264].

The silicon diode, SIT, and ISIT vidicons [261]

The target of the silicon diode vidicon is a two-dimensional array of independent p–n silicon diodes on 15 μm centres. The target also acts as the photosensitive surface. Charge carriers created by the incident light discharge a reverse bias applied through a large load resistor. The electron reading beam completes the circuit to ground. The rebiasing current pulse is sensed by the amplifier to give the video output. The individual diodes have many of the useful properties already outlined for the single diode in Chapter 7, viz. high quantum efficiency (0.8 at 800 nm), large dynamic range (10^6 carriers), and low voltage operation which avoids problems with corona. The target must be refrigerated in accordance with Figure 2.8 for analogue integration.

Fig. 8.16 The technique for finding the centres of photon events and discriminating against ion events with the Boksenberg photon counting system. (Reproduced from [42].)

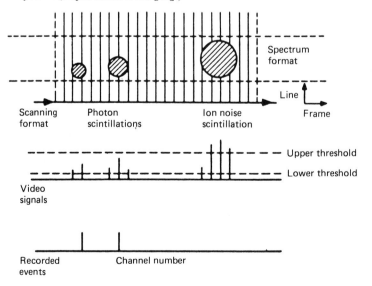

The silicon diode vidicon DQE is severely reduced at low light levels by amplifier and beam noise. The silicon intensified target (SIT) vidicon has a more or less conventional silicon diode vidicon preceded by an electrostatic image intensifier stage. Photoelectrons are detected directly by creating some 10^3 carriers each in the target. This is often known as the electron bombarded silicon (EBS) mode. Since the system readout noise is also about 10^3 electrons r.m.s. the SIT vidicon DQE is close to q, the quantum efficiency of the photocathode material. The spectral and spatial sensitivity is limited by the fibre optic face plate.

An ISIT is an intensified, silicon intensified target vidicon in which a SIT is fibre-optically coupled to an image intensifier for even greater sensitivity.

Image tube scanners (ITS) [329a]

This detector was developed for spectroscopy and uses phosphor persistence as intermediate storage. The phosphor is scanned optically by an image dissector. The latter is a photomultiplier which has a diaphragm or slot before the first dynode. Television type scanning coils deflect and sweep the photoelectron output from the photocathode across the slot. If the scan period is similar to the phosphor scintillation decay time there is a high probability that a scintillation will be detected once and once only. For the ITS in use at the Lick Observatory there are two parallel scans, one for the stellar spectrum plus sky background, the other for sky alone. The total time required to make both scans is 4.3 ms with a total of 4096 individual points at a dwell time of 1 μs per point.

Finding and guiding

Although television cameras have been largely superseded by the solid state devices discussed in the next section they are still used extensively to examine telescope finder fields and for offset guiding [8]. The sensitivity of the camera is normally increased by providing controllable amounts of integration of the image either in the target of the camera tube or offline using a video disc or storage tube [8, 115]. It is also possible to introduce some offline processing for image enhancement and background suppression [400].

Limitations of television cameras

Television cameras suffer from a number of problems which limit their usefulness. These almost all arise from complications in the beam landing and hysteresis in the scan coils. The electrons in the reading beam reach the target with almost zero velocity and as the tip of the beam moves across the target it is easily deflected towards any nearby areas of high positive potential. This causes misregistration in the image reconstruction. The phenomenon is known as *beam pulling* and can cause jitter or instability of up to one or two scan lines.

Only the most energetic electrons in the beam discharge the target areas corresponding to the lowest light levels. This leads to the phenomenon of *lag* in which the high signal areas of the target are discharged to about 10 per cent in a single scan but ten or more scans are required to complete the discharge (indeed, the potentials of many cameras such as the Isocon can be adjusted to give essentially non-destructive readout allowing the target to be scanned up to 10^3 times before they are discharged). When operating in the analogue integration mode a low light level flash together with as many as ten raster scans is usually necessary in order to prepare a target for an exposure. It may also be necessary to turn off the filament to avoid detection of glow during long exposures.

In addition to these problems, vidicons have the large source capacitance of the target associated with the video line. This introduces a large KTC reset noise (see Chapter 7).

Solid state detectors [262, 383]

The use of shift registers with arrays of diodes and the transfer techniques of charge transfer devices (CTD) has provided a range of one- and two-dimensional detectors with reduced source capacitance which do not suffer from any of the above shortcomings, although they have problems of their own.

Electrons and holes generated by photon absorption in the silicon are separated either by a depletion region of a diffused p–n junction or by a field induced junction of a MOS (metal oxide switch) capacitor (CCD, CID).

Self-scanned silicon diode arrays (Reticon) [73, 174, 398, 405, 407]

The operation and characteristics of single silicon diodes have been discussed in Chapter 7. Their most successful use has been in arrays which were developed for control and to read print. Often known as Reticons after one of their manufacturers, these have been used most frequently in a one-dimensional configuration for spectroscopy and consist of a series of p–n junctions having a common gold substrate as anode and overcoated with silicon dioxide. A section of an array is shown in Figure 8.17. The centre-to-centre spacing of the individual diodes which are available in widths of up to 2 mm, can be 15, 25, or 30 μm (100 μm in two-dimensional configurations), and single arrays of up to 4096 diodes (61.5 mm total length) are available. The individual diodes are back biased just like the single diode discussed in Chapter 7 and readout is similar to that of the silicon diode vidicon except that the electron reading beam is replaced by an on-chip shift register.

Figure 8.18 shows, in section, a Reticon mounted in a liquid nitrogen cooled cold box which maintains the array at 93 K. Unavoidable parasitic capacitance between the shift registers and the video line couples the transients of the clock driver signals onto the video line. Figure 8.19(a) shows a section of a stellar spectrum uncorrected for this fixed pattern and Figure 8.19(b) is the same spectrum with the fixed pattern subtracted. The fixed pattern in the baseline can be determined accurately from several short exposures. Such fixed patterns caused by coupling of the clock line transients onto the video line are a common feature of solid state detectors. 'Clean' clock signal waveforms and high stability are crucial if the video signals are to be properly corrected for the fixed pattern.

The silicon in the diode array becomes increasingly transparent for photon energies below the excitation energy particularly when refrigerated. Minority carriers generated more than the diffusion length (10 μm) from the 2 μm thick depletion region are not detected.

Fig. 8.17 Section of a silicon diode array.

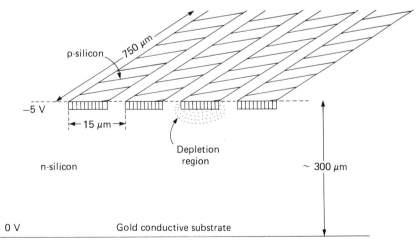

Fig. 8.18 A Reticon mounted in a liquid nitrogen refrigerated housing which can be used in any orientation [73].

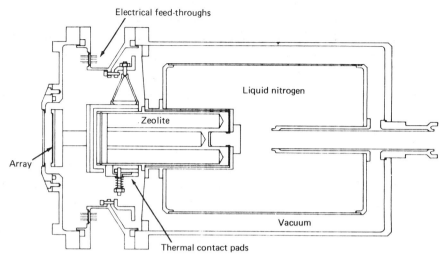

Fig. 8.19 (a) A stellar spectrum taken with a refrigerated Reticon which shows the strong quartet modulation caused by pickup of the clock signals on the video line, (b) the same as (a) but after the baseline has been subtracted and plotted on an expanded scale [407].

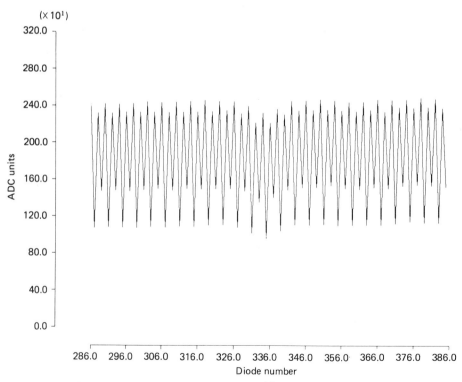

(a)

Transparency of the silicon in the near infrared leads to interference between the incident radiation and that reflected from the gold substrate. In sufficiently monochromatic light this leads to optical fringing. This is a common phenomenon for thin silicon detectors and is discussed below in connection with CCDs. In spectroscopy the fringe amplitude is a function of the free spectral range covered by each diode and by the monochromatic focal ratio of the spectrograph camera.

Analogue integration

The diodes in the array have a common anode which increases the source capacitance on the video line compared with the value for a single diode. It is possible to restrict the reset noise (discussed in Chapter 7) to that appropriate to a single diode (equation (7.11)) by the technique of correlated double sampling (CDS) by using readout and differentially integrating circuits of the kind shown in Figure 8.20 [151, 405, 415]. The technique is also used when reading out CCDs. The source capacitance across the amplifier input, C, is short-circuited by the switch S between the readout of each diode. This means that the 'before' and 'after' voltages for a given diode are measured relative to the same ground reference which,

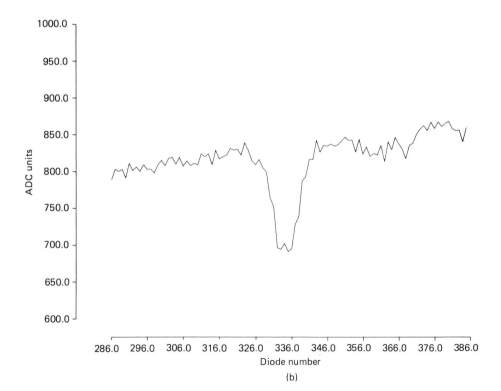

(b)

although it is not stable on long time scales (60 Hz pickup etc.) is relatively constant for the double sample interval of a few microseconds.

The CDS readout sequence for diode 2 is:
   (i) open S and integrate (video line zero reference),
   (ii) open 1, close 2, invert integrator (photo-generated charge),
   (iii) close S (rebias diode 2).

Since reset noise is the principal source of noise it is possible to achieve a system noise as low as 400 electrons (r.m.s.) which corresponds to the smaller stray capacitance, c, across a single diode [398].

The readout noise is still relatively high and clearly makes the device inappropriate at low light levels but the high quantum efficiency and high saturation charge (10^7 electrons) make it particularly efficient for high signal-to-noise (> 100) spectroscopy (see Figure 7.5 and equation (7.14)).

Photon counting

Silicon diodes can either be exposed directly to accelerated photoelectrons [83], or detect phosphor scintillations from an image intensifier [367].

Fig. 8.20 Correlated double sampling circuitry which has the effect of reducing KTC noise to that of a single diode [405].

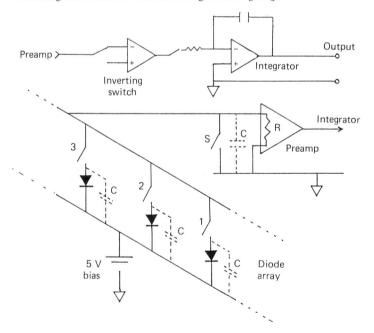

Figure 8.21 shows in block form an intensified Reticon system for the detection of individual photon events [346]. It has some features in common with the TV photon counting system described earlier. There is a photon gain of about 10^8 achieved by fibre-optically coupling two three-stage magnetically focussed image intensifiers. The output of the final phosphor (at $+40\,\text{kV}$) is lens coupled to a diode array (Reticon). The saturation charge for a single diode is about 10^7 electrons in each of three or four adjacent diodes. The array is clocked at $100\,\text{kHz}$ and successive readouts are differenced.

The fixed pattern arising from the pickup of clocking signals on the video line is eliminated by using the difference signal. Photon events are only counted once in the difference signal because it is sensitive only to the rapidly rising leading edge of the scintillation and not to its much slower decay phase.

The diode array has negligible lag below saturation and it is clocked at a much higher rate than the TV camera in the system described earlier. As a result the system has a much higher dynamic range. It is also possible to carry out event centring to improve resolution.

Fig. 8.21 An intensified photon counting Reticon system. (Reproduced from [346].)

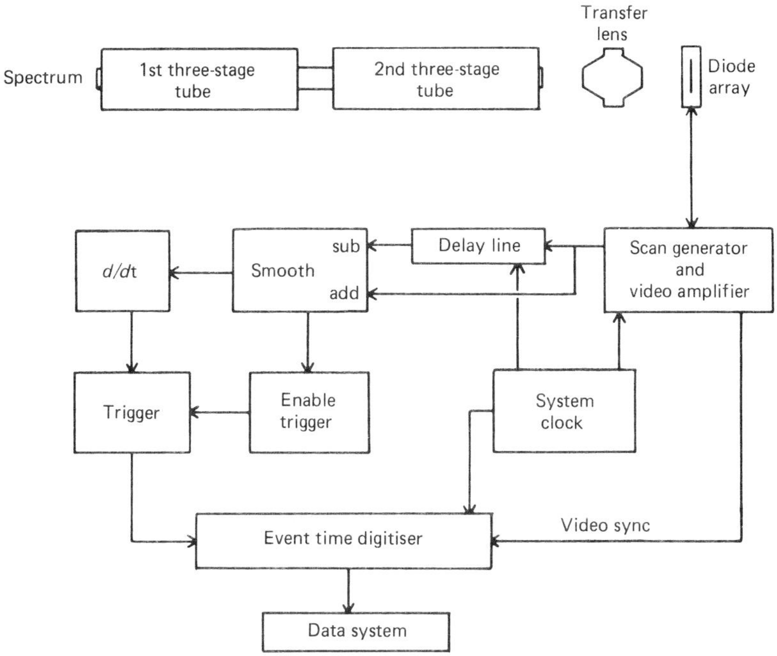

The Digicon [158]

Arrays such as the Reticon have not been very successful in the electron bombarded mode because of electronic damage to the shift registers and surface layers of the diodes, as well as difficulties in processing the photocathode in the same volume [18]. However, the use of arrays of individually connected diodes has produced detectors with excellent photon counting characteristics [158].

Figure 8.22 is a sectional view of a Digicon, which consists of an array of 512 individual silicon diodes each wired to its own charge sensitive preamplifier. The arrival of each photoelectron is registered immediately

Fig. 8.22 Sectional view of a Digicon. (Reproduced from [158].)

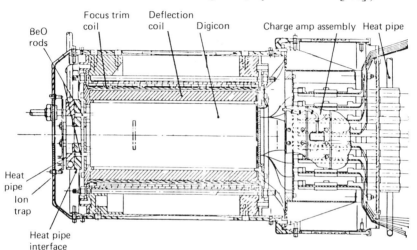

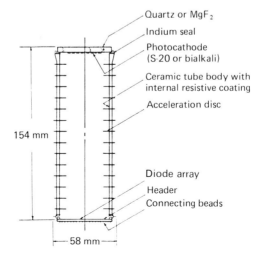

in a memory. There is a magnetic yoke for focussing which also allows deflection or scanning as well as some focus trim. A pair of Digicons will act as the detectors for the Faint Object Spectrograph (FOS) of the Hubble Space Telescope [18].

The detector characteristics are given in Table 8.2. Some 5000 charge carrier pairs are generated for each 22.5 kV electron. The output pulse height distribution (PHD) for a Digicon with an S-20 photocathode is shown in Figure 8.23(a). Figure 8.23(b) shows the response of the same device to a sudden change in incident photon flux (6000 to 1).

The only important shortcomings of the Digicon are the limited number of channels and its rather coarse spatial resolution.

Charge transfer devices

There are two charge transfer devices which have been used for astronomical observations, the charge injection device (CID) [289, 414], and the charge coupled device (CCD) [214, 415]. Both use arrays of MOS (metal oxide switch) capacitors to integrate photogenerated carriers at independent photo sites. The capacitors are biased to create potential wells in the bulk silicon which collect and store the photogenerated electrons.

Table 8.2 *FOS detector requirements*

Dark count rate	$\leqslant 0.002 \, \text{ct s}^{-1}$ at $-10\,°C$
Dynamic range	10^7
Photocathodes	Na_2KSb (bialkali) on MgF_2
	Na_2KSb (Cs) (trialkali) on fused silica
Quantum efficiency	good in terms of current state of the art
	trialkali is similar to S20 response
Photocathode uniformity	$\leqslant \pm 10\%$
Magnification	$1.00 \pm 2\%$
Resolution	limited by $40 \, \mu m$ diode elements
Distortion	$40 \, \mu m$ at 15 mm radius, radial or tangential
Gain	5000 at 20 kV
Photocathode mask	$20 \, \text{mm} \times 30 \, \text{mm}$
Data format used	$7 \, \text{mm} \times 30 \, \text{mm}$
Diode array format	512 diode elements on $50 \, \mu m$ centres
Diode element size	$40 \, \mu m \times 200 \, \mu m$
HV operating range	15 kV to 25 kV
HV current	$\leqslant 18 \, \mu A$ at 22.5 kV
FWHM/en. pk. of PHD	$\leqslant 0.25$

Fig. 8.23 (a) The pulse height distribution of the Digicon with an S-20 photocathode in the Faint Object Spectrograph of the Hubble Space Telescope. (b) The response to a sudden change in photon flux. (Reproduced from [158].)

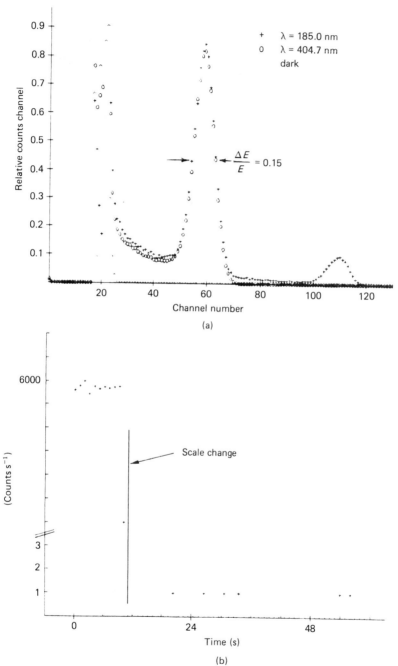

(a)

(b)

Analogue integration

CID [4]

The individual sensing or picture elements of a CID consist of two electrodes. In a two-dimensional array the gates of the capacitors are connected in rows and columns, with one gate from each sensing element forming the rows while the others form the columns, as is shown schematically in Figure 8.24. The charge accumulated at any picture element can be sensed non-destructively as shown in Figure 8.24 by changing the potentials on the gates to move the charge back and forth between the two capacitors and sensing the voltage change on one of them. The voltage difference is simply the charge divided by the column capacity.

Although CIDs have been used for astronomical observations they suffer from some low light level reciprocity failure and a higher reset noise than a CCD because the gates within a column are connected in parallel [4]. The non-destructive readout capability of the CID means that the reset noise can be reduced by measuring the charge a large number of times (up to 10^3). Reciprocity failure can be overcome by the use of a low light level flash.

Fig. 8.24 Readout scheme of a CID array where ΔV is the signal charge divided by the column capacity.

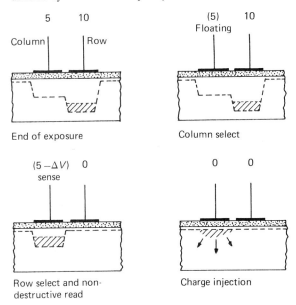

CCD [214, 415]

Charge coupled devices are used extensively as detectors in astronomy. They are sensitive from the X-ray region to the long wavelength cutoff of the excitation energy and the r.m.s. readout noise of some devices is less than ten equivalent photons. The introduction of hybrid, tri-metal or InSb arrays mated (bump bonded) to a CCD as multiplexer has extended the range of sensitivity to the 1 to 15 μm range [141].

Figure 8.25 shows a schematic of a CCD area photometer in which the device can be used directly or behind an image intensifier [199]. At the end of an exposure, charge is shifted successively to the on-chip amplifier and sampled and digitised by the associated electronics. The data is both stored and displayed using a small computer.

The capacitors of the CCD are arranged in rows and columns. At the end of an exposure the charge packets which have accumulated under the capacitor gates in each row are transferred along the columns successively to the subsequent rows. The charges in the final row are delivered individually to the on-chip amplifier. The technique, sometimes known as a bucket brigade, is illustrated in Figure 8.26. It limits the source capacitance to < 0.01 pF and, consequently, reset noise is only a few tens of electrons (r.m.s.) or less. The saturation charge or full well capacity is about 10^5 electrons.

Only a few CCDs are suitable for astronomy and some of those available at the time of writing are listed in Table 8.3 together with their

Fig. 8.25 Block diagram of a typical CCD area photometric system. (Published with permission from [199].)

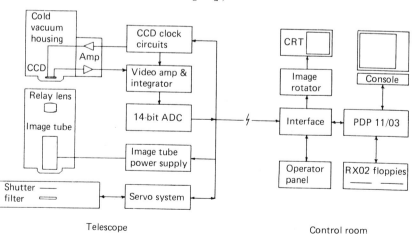

characteristics [38, 143, 167, 199, 266, 408]. They differ from each other in electrode configuration, thickness, number of sensing elements, and spectral response. In so-called front illuminated devices the light passes through the electrodes before reaching the silicon. In rear, or backside, illuminated devices the silicon is thinned to about 10 μm which allows the light to go directly into the silicon without passing through the electrodes.

Both front and rear illuminated devices suffer from optical interference effects, in one case because of the thin, multiple electrode structure, and in the other because of the thin silicon layer. In a few CCDs the size of the photosensitive elements is limited by dead space (insensitive to light). In this case particular care must be taken to see that the sampling frequency of the point spread function is high enough to avoid the aliasing problems discussed in Chapter 4.

Figure 8.27 shows how the photon absorption length in silicon varies with wavelength of the incident radiation. For wavelengths shorter than

Fig. 8.26 A three-phase CCD charge transfer system schematic.

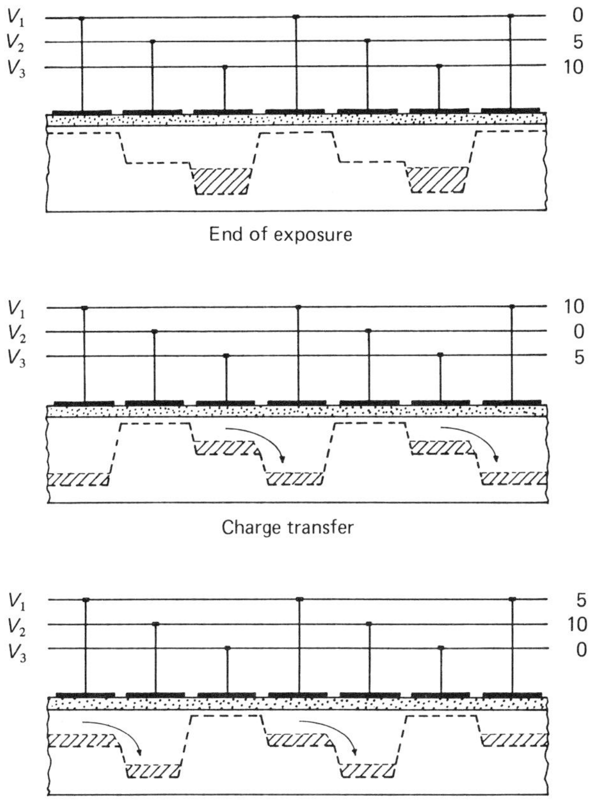

End of exposure

Charge transfer

Table 8.3 *CCDs suitable for astronomy* (1986)

Device	Pixel size (μm)	Format	Peak QE	R.m.s. noise (e^-)	Status comments
General Electric Company (UK)					
GEC P8603A	22	385×576	0.3	15	available
Thomson CSF (France)					
THX31133	23	384×576	0.3	23	available
Tektronix					
512×512	27	512×512	$0.8 + UV$	10	mid-1986
2048×2048	27	2048×2048	$0.8 + UV$	10	1987?
RCA					
SID-501	30	320×512	0.8	60	no longer available
SID006EX007	15	640×1024	$0.8 + UV$	50	no longer available
Texas Instruments					
4849	22	390×584	0.4	100	not available outside USA
800×800	15	800×800	0.8	7	limited quantity
1024×1024	18	1024×1024	0.5	50	

400 nm, carriers are generated very close (<10 nm) to the surface [214]. Consequently, the backside illuminated devices have a much better potential for high ultraviolet sensitivity than the thick, front illuminated devices.

The silicon has a natural coating of silicon dioxide and is often bonded to a glass cover slip. Although the oxide and glass are essentially transparent, positive charge trapped at the silicon/silicon dioxide boundary creates a potential well which can trap electrons generated near the silicon surface. If the silicon layer is too thin, electrons tend to recombine with trapped holes. If the layer is too thick UV photons are

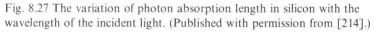

Fig. 8.27 The variation of photon absorption length in silicon with the wavelength of the incident light. (Published with permission from [214].)

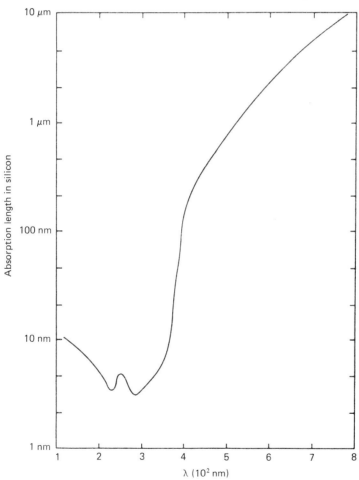

absorbed before reaching the region of the potential well beneath the capacitor gates. In either case the blue/ultraviolet sensitivity is reduced.

The stability, uniformity and the size of the blue/UV sensitivity depends on the degree to which the silicon thickness can be controlled in manufacture and the use of such techniques as ion implantation, or UV (<250 nm) flooding in the presence of oxygen or nitrous oxide to negatively charge the backside to reduce the depth of the surface well. Figure 8.28 shows some typical CCD spectral sensitivity curves.

The variation in sensitivity from photo site to photo site is about 5 per cent and this can be calibrated by exposure to a uniformly illuminated, or 'flat', field such as the dawn sky, a dark globule in the Milky Way, or the illuminated interior of the telescope dome, although none of these is perfectly uniform or unpolarised [24, 198].

The effect of element-to-element sensitivity variations can be minimised by a technique known as time delay and integrate (TDI), or drift scan [424]. The photometer is mounted on a carriage which drives the CCD parallel to the direction of its columns. The rows of the device are clocked to the on-chip amplifier at a rate which just matches the drift of the image

Fig. 8.28 Spectral sensitivity curves for various thinned and unthinned CCDs.

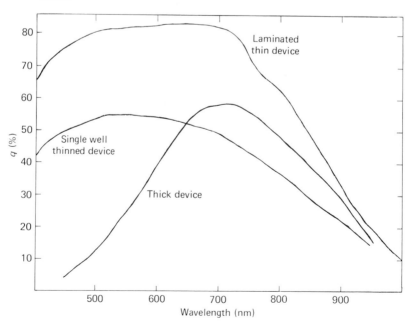

across the CCD. In effect this means that a charge packet tracks its associated element in the image with some charge being generated at each photosite in the column and the photosite response is the average for all of the photosites in a column.

The TDI technique is successful as long as the sky or background contribution remains roughly constant during the exposure. By moving the detector rather than allowing the telescope to drift in hour angle, the nonlinear motions which would result from optical distortions of the field (particularly for a flat focal surface) are avoided. Further, since the motions of stars are small circles on the sky there would be significant differential motion across the field if the stars were allowed to drift in hour angle. This would produce elongated images unless some compensating variation in magnification were provided across the field.

If the charge transfer efficiency between photosites is not > 99.9999 per cent then serious 'smearing' of charge will take place which is particularly obvious for bright stars. Charge transfer inefficiency is particularly serious for some CCDs when the amount of charge in individual charge packets is similar to the on-chip amplifier r.m.s. readout noise. Figure 8.29 shows how the transfer efficiency along a column was found to vary with signal level for a single transfer of charge for an older, unthinned RCA device

Fig. 8.29 The variation of charge transfer efficiency with signal level for a single transfer of charge along a column as measured for an RCA CCD. 1 ADCU (analogue to digital converter unit) is equivalent to 27 detected photons [408].

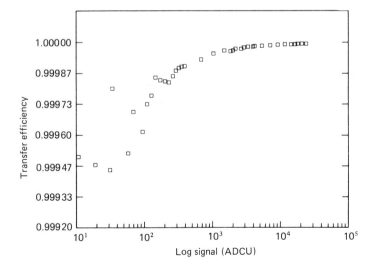

[408]. The calibration of ADCU (analogue to digital converter units) is about 27 electrons (detected photons) per unit and the r.m.s. readout noise is 70 electrons. The loss of transfer efficiency at low signal levels is quite severe. The effect is less marked in more recent devices.

If a site contains a charge Q and a fraction $f(Q)$ is transferred to the adjacent site at a single transfer then the transfer efficiency is $f(Q)/Q$. Empirically, $f(Q)$ is found to have the form

$$1 - f(Q) = \text{constant} \cdot Q^{0.5} \tag{8.17}$$

with the constant being approximately unity [198]. Both the constant and the power are device specific. The function $(f(Q))^{-1}$ can be used at high values of Q to restore an image but it is not effective at low values because of noise.

Optical fringing is caused by interference in the thin silicon layer, and in some cases in the transparent electrodes, while the contours of the fringes are affected by irregularities in the thickness of the silicon. The effect depends both on the wavelength and on the band width of the radiation. Silicon becomes increasingly transparent at longer wavelengths (see Figure 8.27) which increases the proportion of radiation reflected from the rear surface thereby increasing the fringe modulation. Figure 8.30 shows a particularly striking example of fringing in a CCD image of the Orion nebula taken in the effectively monochromatic light of the [S III] ion at

Fig. 8.30 A CCD image of the Orion Nebula in the light of [S III] (at 953.2 nm) which shows marked fringing and two bad columns [129].

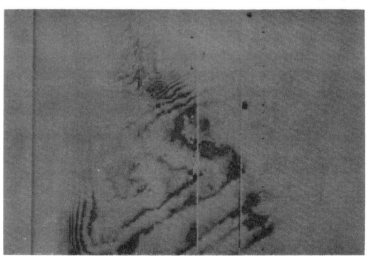

953.2 nm [129]. To date, internal calibration through the use of a 'flat' field seems to adequately correct for fringing in broad band observations but not for monochromatic observations of emission line objects or night sky emission lines in the red/infrared.

Photon counting

CCDs operating in the electron bombarded mode (ICCD) within an image intensifier are being developed for use in the rocket ultraviolet and optical spectral regions [126, 260, 367].

CCDs are used in two-dimensional systems to detect and register phosphor scintillations caused by individual photon events in high gain image intensifiers [368]. The technique is analogous to the TV photon counting systems described above. Event centring is possible when the array is fibre-optically coupled to the intensifier and the photon gain is $> 10^6$ or, 10^7 if it is lens coupled. A single stage, curved channel MCP or a two-stage chevron channel MCP gives sufficient gain for fibre optic coupling. To avoid optical cross talk in continuous operation a CCD such as the Fairchild 222A is preferred in which charge from each pixel is moved behind an occulting aluminium strip before transfer along the column to the on-chip amplifier. This assumes that the scintillation is sampled by several CCD pixels.

There are several advantages to using a CCD in such a photon counting mode. The array operates continuously and since dark current is unimportant it does not need to be refrigerated. It is possible to continuously monitor the signal and cosmic ray events can be eliminated (see discussion below). The principal restrictions are the lower quantum efficiency of semitransparent photocathodes in the visible spectrum, the dynamic range which is limited to a few hertz per pixel, and the complicated photometric response especially when a MCP input is used. These last two points make it difficult to calibrate photometrically.

Multi-anode, resistive anode, and microchannel arrays
[245, 383]

These devices register the electron bursts from microchannel arrays with a position sensitive anode without intermediate storage in a phosphor or a detector such as a CCD.

Resistive anode MCP

In its simplest form four amplifiers are attached diagonally to a resistive anode and the position of the incident charge pulse is calculated from the division of charge detected at each amplifier. Unfortunately a regular circular or square geometry for the anode leads to image distortion in the reconstruction.

Figure 8.31 shows the geometry of a circular arc terminated resistive anode which produces less than 5 per cent linear distortion in the image [243]. Charge sensitive amplifiers are attached at each of the four corners. If the sheet resistivity of the anode is r (usually a few times 10^5 ohms) then the shunt resistance, $R(L)$, at the border of the anode has the value

$$R(L)=r/a \quad \Omega \, \text{mm}^{-1} \tag{8.18}$$

where a is the radius of curvature of the edge (as shown) in millimetres. The cartesian coordinates (x, y) of the initial, incident, charge pulse are given by

$$x/d=(I_1+I_4)/(I_1+I_2+I_3+I_4) \tag{8.19}$$
$$y/d=(I_1+I_2)/(I_1+I_2+I_3+I_4) \tag{8.20}$$

where d is length of the chord between corners (as shown) and I_n is the current detected on amplifier n. For this configuration the spatial resolution is uniform over the anode area and is limited by thermal noise in the anode. The latter is anticorrelated between opposite sides of the anode and therefore contributes only to the numerators in equations (8.19) and (8.20).

Fig. 8.31 The geometry of a circular arc terminated resistive anode. Electrical connections are made at the corners. (Reproduced from [243].)

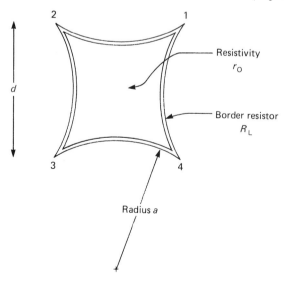

A one-dimensional resistive anode with optimally matched amplifier time constants has a thermal noise which is proportional to anode capacitance, while a circular arc terminated anode of the same capacitance has a mean square noise which is approximately $(1 + 2a/d)$ times greater. A typical resolution is 50 per cent modulation at 10 line pairs per millimetre over a 25 mm diameter anode with a dual microchannel plate giving an electron gain of 10^7 where the thermal noise is equivalent to about 6×10^{-16} C in 1 μs [243].

The microchannel array (MAMA) detector [383, 384]

This detector has been developed as a compact, rugged, device for ultraviolet observations from rockets. The anode is proximity focussed to the MCP output. There are two anode forms, discrete and coincidence. In the discrete form there is an array of separate anodes each with its own charge sensitive amplifier and discriminator circuit. The total number of possible anode elements is restricted to about 500.

The greatest interest for astronomy lies in the coincidence anode design, an example of which is shown in Figure 8.32. There are two orthogonal grids of electrodes with 25 μm centre-to-centre spacing in each grid. Each electrode is attached to a charge sensitive amplifier and discriminator circuit which issues a logic pulse for a MCP charge pulse which exceeds a preset threshold. If the numbers of electrodes on the two grids are a and b then there are $a \times b$ picture elements but only $a + b$ amplifiers. A pulse position is determined by simultaneous detection on two electrodes when the appropriate memory location of the associated data acquisition system is incremented by one.

The maximum decoding and storage rate is set by the pulse pair resolution of the electronics which is of the order of 1 μs. Although count rates as high as 10^6 Hz are possible from the whole array it can be overloaded at much lower rates if light is concentrated in one area by a bright star image. A high flux of pulses from one star would tend to produce 'ghost' images through pulse coincidences with other faint stars in the field.

Effect of cosmic rays [3, 143, 250, 276, 430]

The passage of cosmic ray secondaries or radioactive decay products through incoherent detectors is an important source of noise when they operate in an integrating mode at low signal levels. Cosmic ray muons have typical energies of 400 MeV and are extremely penetrating suffering little attenuation in their passage through the atmosphere or observatory buildings. They deposit about $1\,\text{MeV gm}^{-1}\,\text{cm}^{-2}$ when passing through silicon which, for an electron diffusion length of $30\,\mu\text{m}$, means about 2000 electrons/passage [3]. For the much thinner photocathode layers the number is proportionately smaller.

Fig. 8.32 A 512×512 coincidence anode array used with a MAMA detector. (Published with permission from [384].)

The spike of charge is confined to a few pixels in the integrating detector or to a large, single photon event for a photon counting system. The latter can be eliminated by setting an upper discriminator level for detection. An example of the effect on a weak spectrum taken with an integrating Reticon detector is shown in Figure 8.33. Details of the exposure are given in the caption. In thick devices the events can be recognised as trails through several photosites, while in thinned devices an event tends to be concentrated in a single pixel [143].

The frequency of cosmic ray muon events in silicon detectors is approximately $1 \, h^{-1} \, mm^{-2}$ at 5000 m altitude and $0.5 \, h^{-1} \, mm^{-2}$ at sea level. The number can fluctuate by a factor of two depending on the level of solar activity. The rare passage of short range α-particles produces massive spikes of some 10^5 electrons amplitude. Such a double event is shown in Figure 8.33. These large events occur about once per month per square millimetre. Similar effects also arise from the daughter products of radioactive decay of any radioactive contamination of the detector packaging [176, 430].

Photon counting vs. integration for visible detectors

It is useful to compare the performance of photon counting and integrating detectors in the two critical areas of (i) sky-limited photometry, (ii) sky-limited spectroscopy/narrow band photometry.

In (i), detector noise is less important than sky noise, and here the important difference between the systems is the higher quantum efficiency of the solid state sensors compared with semitransparent photocathodes of the intensified systems and the low maximum count rate (about 1 Hz) of the multiplexed photon counting systems. This is illustrated in Figure 8.34 for exposures with a 3.6 m telescope where $\log(s)$, the signal-to-noise, is plotted against stellar magnitude for a photon counting detector with an S-20 photocathode and a CCD both operating close to the wavelengths of their respective peak quantum efficiencies.

The calculations are based on equation (2.18) in Chapter 2 with the same assumptions concerning image quality. The exposure time is 1133 seconds and the filter bandpass is 50 nm. The values of the variables used in equation (2.18) are $q = 0.8$, $D = 400$, and $m = 5$ for the CCD, and $q = 0.15$ and $D = 0$ for the photon counting system. The sky brightness of 22.5 magnitude per square arc second corresponds to 30 photons s^{-1} $arcsec^{-2}$.

The turnover in the curves at bright magnitudes is caused by pile-up

Fig. 8.33 Cosmic ray spikes in spectra taken in 6000 second exposures with an integrating Reticon detector at Mauna Kea, Hawaii (5000 m). The spikes of about 1500 electrons in the upper spectrum are probably due to muons, while the spikes of about 10^5 electrons in the lower spectrum are probably caused by alpha particles.

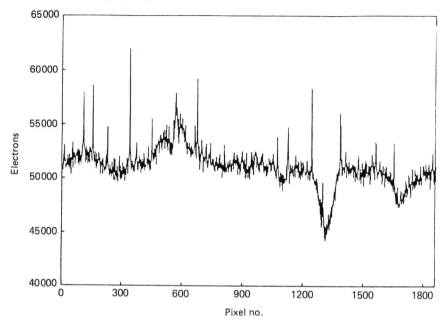

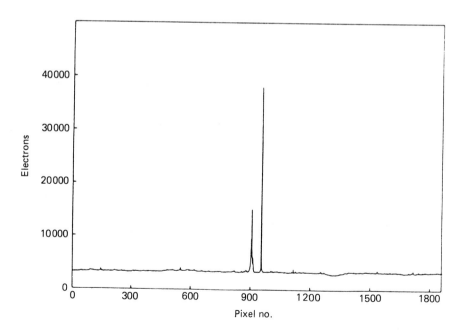

errors in the photon counting system and saturation in the case of the CCD. To match the performance of the CCD the photon counting system would have to be exposed six times longer (i.e. the ratio of the quantum efficiencies). In addition the CCD has a wider dynamic range.

Figure 8.35 illustrates the real advantage of the photon counting system at the low signal levels typical of (ii) where the balance between sky and detector noise depends critically on exposure time. The limiting stellar magnitude for a signal/noise = 5 (typical for detection of features in quasi-stellar object spectra) is shown as a function of exposure time for three systems. The sky brightness is again taken as 22.5 magnitude per square arc second but the bandpass is only 0.5 nm. The dotted line is for a photon counting system with the same characteristics as above. It saturates for sources brighter than 19 magnitude. The lower, solid line is for the same CCD as above. The upper solid line is for the same CCD except that in this case the charge packets delivered to the on-chip amplifier have been binned normal to the spectrum by appropriate clocking of the CCD. The

Fig. 8.34 Variation of the logarithm of signal-to-noise (s) with stellar magnitude for broad band photometry (50 nm filter) with a 3.6 m telescope for a photon counting system (dots) and an integrating CCD system (solid line). The sky brightness is 22.5 mag arcsec^{-2} and the exposure time is 1133 seconds.

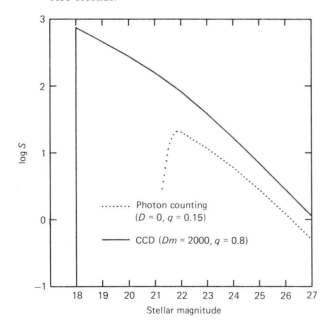

contributions of 2 or more columns have been added to give a single spectrum of the source such that $m = 2.5$. This reduces the contribution of the readout noise while still properly sampling the LSF.

Although from Figure 8.35 the CCDs are both better than the photon counting system at long exposures where the sky noise dominates and the binned CCD is superior at all exposure times, they suffer from a serious disadvantage. It is not possible to monitor the signal of the CCDs during the exposure. Also, binning presupposes knowledge of the exact position of the source on the slit. These are serious disadvantages for unknown objects. To calculate the effective readout noise of the CCD system, *Dm* must be multiplied by the number of times it is read out to monitor the exposure. Even for two readouts the binned CCD system would have an inferior performance to the photon counting system.

The situation is more extreme in the few-photon case (in both rate and total) of, say, multi-filter photometry, where many short exposures through a series of filters are accumulated. The reason for taking many cycles is to limit the effect of changing sky conditions. Figure 8.36 is a plot of $\log(s)$ against $\log(N)$, the total number of incident photons in an exposure. The two lower solid lines are for the binned and unbinned CCDs. The lower dotted line is for the photon counting system already

Fig. 8.35 The limiting stellar magnitude detected with a signal/noise = 5 as a function of exposure time with a filter of 0.5 nm width on a 3.6 m telescope for a photon counting system and a CCD system with $D^{1/2} = 10$, $m = 2.5$ (binned) and a CCD system with $D^{1/2} = 20$, $m = 5$ (unbinned). See text for further details.

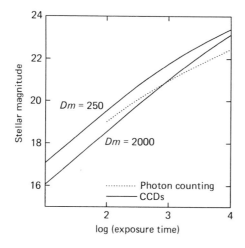

discussed, while the upper is for a system with $q=0.8$, the same quantum efficiency as silicon. In this comparison, the readout noise of the CCD is dominant and, for fewer than 100 incident photons, the photon counting system is superior.

CCDs with $D^{0.5}<10$ are expected commercially soon. The TI CCDs on the WFPC of the Hubble Space Telescope already approach this figure. As $D^{0.5}$ is reduced the lines for the CCDs in Figure 8.36 approach the ideal photon counting system, as shown by the dashed line for binned data with $D^{0.5}=3$, i.e. $Dm=18$.

High resolution imaging X-ray detector (HRI) [208, 225]

This highly successful detector has been used in the 0.15 to 3 keV range on its own and in combination with the broad band filter spectrometer (BBFS) and the objective grating spectrometer (OGS) of the Einstein (HEAO 2) satellite observatory (discussed in Chapter 3) [153]. The principle of operation is similar to that of the MAMA and resistive anode detectors described above.

Figure 8.37 shows a schematic of the HRI. There are two MCPs in cascade, separated by $38\,\mu m$, with the first having straight channels perpendicular to the face plate and the second having straight channels

Fig. 8.36 The variation of the detected signal-to-noise, S, with the number of photons, N, incident on the detector for the photon counting system and binned and unbinned systems of Figure 8.35 and an ideal photon counting system with $q=0.8$, and a future CCD with $Dm=18$ (binned).

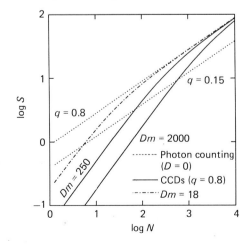

inclined at 6° to the face. In both cases the centre-to-centre spacing of the channels is 16 μm. The anode grids consist of 100 μm diameter molybdenum wires on 200 μm centres. The wires are connected to a chain of 10 kΩ resistors (as shown) with every eighth wire connected to a charge sensitive amplifier. There is a metallic reflector plate behind the anode array.

The X-rays enter the first MCP at angles between 2.5° and 5° as a result of the grazing incidence geometry of the telescope optics (see Figure 3.25) [292, 395]. The input face of the MCP is coated with magnesium fluoride to improve the efficiency of photoelectron production giving a quantum efficiency of about 10 per cent. As a result of the separation between the MCPs the output pulse from the second MCP is spread over several channels and the charge is spread over several of the anode wires.

Coarse determination of pulse position is given by the relative amplitudes of the amplifier outputs while fine position information comes from the relative rise times of the pulses at each amplifier. If a pulse is centred on one of the wires attached to an amplifier a fast pulse of some 10 ns width is detected. A more slowly rising pulse would be detected by the amplifiers on either side with the degradation being nRC, where n is the number of intervening resistors of resistance R, and C is the dynamic input capacitance of the amplifier.

Fig. 8.37 Principal components of the High Resolution Imaging X-ray Detector of the Einstein satellite observatory. (Reproduced from [225].)

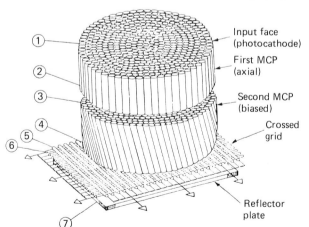

(Not to scale)

APPENDIX A

SI (Le Système International d'Unités)
units and symbols

Physical quantity	Unit	Unit symbol
length	metre	m
mass	kilogram	kg
time	second	s
electric current	ampere	A
thermodynamic temperature	kelvin	K
plane angle	radian	rad
solid angle	steradian	sr

Physical quantity	Unit	Definition of unit	Unit symbol
energy	joule	$m^2\,kg\,s^{-2}$	J
pressure	pascal	$kg\,m^{-1}\,s^{-2}$	Pa
power	watt	$m^2\,kg\,s^{-3}$	W
electric charge	coulomb	$A\,s$	C
electric potential difference	volt	$m^2\,kg\,s^{-3}\,A^{-1}$	V
electric resistance	ohm	$m^2\,kg\,s^{-3}\,A^{-2}$	Ω
electric capacitance	farad	$s^4\,A^2\,m^{-2}\,kg^{-1}$	F
frequency	hertz	s^{-1}	Hz
magnetic flux density	tesla	$kg\,s^{-2}\,A^{-1}$	T

SI prefixes

Multiple	Prefix	Symbol	Multiple	Prefix	Symbol
10^{-1}	deci	d	10	deca	da
10^{-2}	centi	c	10^2	hecto	h
10^{-3}	milli	m	10^3	kilo	k
10^{-6}	micro	μ	10^6	mega	M
10^{-9}	nano	n	10^9	giga	G
10^{-12}	pico	p	10^{12}	tera	T
10^{-15}	femto	f	10^{15}	peta	P
10^{-18}	atto	a	10^{18}	exa	E

Conversions to SI

Unit	Unit symbol	SI equivalent
1 electron-volt	eV	1.602×10^{-19} J
1 erg	erg	10^{-7} J
1 jansky	Jy	10^{-26} W m^{-2} Hz^{-1}
1 ångström	Å	10^{-10} m
1 gauss	G	10^{-4} T
1 atmosphere	atm	101 325 Pa
1 torr	Torr	133.3 Pa
1 astronomical unit	AU	1.496×10^{11} m
1 degree of arc	$^\circ$	1.745×10^{-2} rad
1 minute of arc	′ (arcmin)	2.909×10^{-4} rad
1 second of arc	″ (arcsec)	4.848×10^{-6} rad

APPENDIX B

Errors of observation

The errors associated with the measurement of a physical quantity are of two types.

(1) *Systematic errors* are usually caused by inaccurate calibration or consistent errors on the part of the observer. This type of error is harder to detect by repeated observations with the same equipment.

(2) *Random errors* are due to limitations of the equipment, and shot and wave noise associated with the detected source flux. These errors can be reduced by combining many measurements.

Random errors are generally described by the following quantities. The *mean value, M,* of a set of measurements is

$$M = (X_1 + X_2 + X_3 + \cdots + X_n)/n = \sum X_i/n$$

where n is the total number of observations recorded.

The difference of each value from the mean, $(X_i - M)$, is known as the deviation. The *standard deviation, σ,* is defined as the dispersion from the mean, where

$$\sigma = \{\sum (X_i - M)^2/(n-1)\}^{0.5}$$

Often it is important to determine the accuracy of the mean for a given set of observations. This is represented by the *standard error of the mean,* α, where

$$\alpha = \sigma/n^{0.5} = \{\sum (X_i - M)^2/(n(n-1))\}^{0.5}$$

The accuracy of the mean increases as the square root of the number of observations.

Normal (or Gaussian) distributions are important in the representation of random errors. The normal error curve is shown in Figure B1 and is described by the following equation

$$y = 1/(\sigma(2\pi)^{0.5}) \cdot \exp(-(X - M)^2/(2\sigma^2))$$

y has its maximum value $(= 1/(\sigma(2\pi)^{0.5}))$ when $X = M$.

When $X = M - \sigma$, or $M + \sigma$,

$$y = y_{max} e^{-0.5} = 0.607 y_{max}$$

The area under the curve for values of X between $M - \sigma$ and $M + \sigma$ is 68.27 per cent of the total, and it increases to 95.43 per cent for values between $M - 2\sigma$ and $M + 2\sigma$, and to 99.73 per cent between $M - 3\sigma$ and $M + 3\sigma$.

Fig. B.1 A normal (or Gaussian) error distribution.

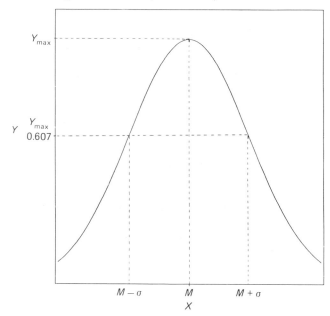

REFERENCES

1. Aarons, J., Guidice, D. A., 1966, The size of low-latitude ionospheric irregularities determined from observations of discrete sources of different angular diameters, *J. Geophys. Res.*, **71**, p. 3277.

2. Ables, J. G., Cooper, B. F. C., Hunt, A. J., Moorey, G. G., Brooks, J. W., 1975, A 1024-channel digital correlator, *Rev. Sci. Instr.*, **46**, p. 284.

3. Aguilar-Benitez, M., Cahn, R. N., Crawford, R. L., Frosch, R., Gopal, G. P., Hendrick, R. E., Hernandez, J. J., Hohler, G., Losty, M. J., Montanet, L., Porter, F. C., Rittenberg, A., Roos, M., Roper, L. D., Shimanda, T., Shrock, R. E., Tornqvist, N. A., Trippe, T. G., Trower, W. P., Walck, Ch., Wohl, C. G., Yost, G. P., Armstrong, B., 1984, Review of particle properties, *Rev. Mod. Phys.*, **56**.

4. Aikens, R. S., Harvey, J. W., Lynds, C. R., 1976, The Kitt Peak CID systems, *IAU Coll. 40*, ed. Duchesne and Lelièvre, Obs. Paris-Meudon, p. 25-1.

5. Allen, C. W., 1976, *Astrophysical Quantities*, Univ. of London, Athlone Press.

6. Anandarao, B. G., Wijnbergen, J., Lena, P., 1983, A scanning metallic-mesh Fabry–Perot interferometer for airborne infrared astronomy, *SPIE*, **445**, p. 42.

7. Angel, J. R. P., Adams, M. T., Boronson, T. A., Moore, R. L., 1977, A very large optical telescope array linked with fused silica fibers, *Astrophys. J.*, **218**, p. 776.

8. Angel, J. R. P., Cromwell, R. H., Magner, J., 1979, An intensified storage vidicon camera for finding and guiding at the telescope, *Adv. Electron. Electron Phys.*, **52**, p. 347.

9. Angel, J. R. P., 1982, Very large ground-based telescopes for optical and ir astronomy, *Nature*, **295**, p. 651.

10. Atherton, P. D., Taylor, K., Pike, C. D., Harmer, C. F. W., Parker, N. M., Hook, R. N., 1982, Taurus: a wide-field imaging Fabry–Perot spectrometer for astronomy, *Mon. Not. R. astron. Soc.*, **201**, p. 661.

11. Aumann, H. H., Walker, R. G., 1977, Infrared astronomical satellite, *Opt. Eng.*, **16**, p. 537.

12. Baars, J. W. M., van der Brugge, J. F., Casse, J. L., Hamaker, J. P., Sondaar, L. H., Visser, J. J., Wellington, K. J., 1973, The synthesis radio telescope at Westerbork, *IEEE Trans.*, **61**, p. 1258.

13. Babcock, H. W., 1960, Stellar magnetic fields, *Stellar Atmospheres*, ed. Greenstein, Univ. of Chicago Press, p. 282.

14. Babcock, H. W., 1962, Measurement of stellar magnetic fields, *Astronomical Techniques*, ed. Hiltner, Univ. of Chicago Press, p. 107.

15.   Babcock, T. A., 1976, A review of methods and mechanisms of hypersensitisation, *AAS Photo-Bull.*, **13**, p. 3.

16.   Backer, D. C., 1981, USA VLBI network handbook, available from Radio Astronomy Lab., Univ. of California, Berkeley, CA 947290.

17.   Baez, A. V., 1960, A proposed X-ray telescope for the 1 to 100 A region, *J. Geophys. Res.*, **65**, p. 3019.

18.   Bahcall, J. N., O'Dell, C. R., 1979, The space telescope observatory, *IAU Coll. 54*, NASA CP-2111, p. 5.

19.   Baldwin, J. E., Field, C., Warner, P. J., Wright, M. C. H., 1971, Mapping of neutral hydrogen in galaxies by aperture synthesis techniques, *Mon. Not. R. astron. Soc.*, **154**, p. 445.

20.   Barden, S. C., Ramsey, L. W., Truax, R. J., 1981, Evaluation of some fiber optical waveguides for astronomical instrumentation, *Pub. astron. Soc. Pacific*, **93**, p. 154.

21.   Bare, C., Clark, B. G., Kellerman, K. I., Cohen, M. H., Jauncey, D. L., 1967, Interferometer experiment with independent local oscillators, *Science*, **157**, p. 189.

22.   Barlow, B. V., 1975, *The Astronomical Telescope*, Wykeham Publications.

23.   Barrow, W. L., Chu, L. J., 1939, Theory of the electromagnetic horn, *Proc. Inst. Radio Engrs.*, **27**, p. 51.

24.   Baum, W. A., Thomsen, B., Kreidl, T. J., 1981, Subtleties in the flat-fielding of charge-coupled device (CCD) images, *SPIE*, **290**, p. 24.

25.   Baustian, W. W., 1960, The Lick Observatory 120-inch telescope, *Telescopes*, ed. Kuiper and Middlehurst, Univ. of Chicago Press, p. 16.

26.   Beardsley, T., 1985, Caltech takes Texan wind, *Nature*, **314**, p. 571.

27.   Beckers, J. M., Hege, E. K., Murphy, H. P., 1983, The differential speckle interferometer, *SPIE*, **445**, p. 462.

28.   Bennett, K., Bignami, G. F., Boella, G., Buccheri, R., Gorisse, M., Hermsen, W., Kanbach, G., Lichti, G. G., Mayer-Hasselwander, H. A., Paul, J. A., Scarsi, L., Swanenburg, B. N., Taylor, B. G., Wills, R. D., 1976, Cos-B observation of high-energy gamma radiation from the Vela region, *Astron. Astrophys.*, **50**, p. 157.

29.   Bernacca, P. L., 1979, The astrometry mission, *Colloquium on European Satellite Astrometry*, ed. Barbieri and Bernacca, TIP Antoniana, Padova, Italy, p. 16.

30.   Bertolli, M., Carnevale, M., Muzii, L., Sette, D., 1968, Interferometric study of phase fluctuations of a laser beam through the turbulent atmosphere, *App. Optics*, **7**, p. 2246.

31.   Betz, A. L., Johnson, M. A., McLaren, R. A., Sutton, E. C., 1976, Heterodyne detection of CO_2 emission lines and wind velocities in the atmosphere of Venus, *Astrophys. J.*, **208**, p. L141.

32.   Betz, A., 1980, Infrared heterodyne spectroscopy in astronomy, *NASA Conf. Publ.*, **2138**, p. 11.

33.   Bignami, G. F., Hermsen, W., 1983, Galactic gamma-ray sources, *A. Rev. Astron. Astrophys.*, **21**, p. 67.

34.   Bingham, R. G., 1979, Grating spectrometers re-examined, *Q. J. R. astron. Soc.*, **20**, p. 395.

35.   Bisnovaty-Kogan, G. S., Estulin, I. V., Havenson, N. G., Kurt, V. G., Mersov, G. A., Novikov, I. D., 1981, The transformation of the coordinates of a gamma burst source to a star catalogue, *Astrophys. Space Sci.*, **75**, p. 219.

36. Black, D. C. (ed.), 1980, *Project Orion (a Design Study of a System for Detecting Extrasolar Planets)*, NASA SP-436.

37. Blackwell, D. E., Shallis, M. J., 1977, Stellar angular diameters from infrared photometry. Application to Arcturus and other stars with effective temperatures, *Mon. Not. R. astron. Soc.*, **180**, p. 177.

38. Blouke, M. M., Janesick, J. R., Hall, J. E., Cownes, M. W., 1981, Texas Instruments (TI) charge-coupled device (CCD) image sensor, *SPIE*, **290**, p. 6.

39. Blythe, J. H., 1957, A new type of pencil-beam aerial for radio astronomy, *Mon. Not. R. astron. Soc.*, **117**, p. 644.

40. Bogess, A., Carr, F. A., Evans, D. C., Fischel, D., Freeman, H. R., Fuechsel, C. F., Klinglesmith, D. A., Krueger, V. L., Longanecker, G. W., Moore, J. V., Pyle, E. J., Rebar, F., Sizemore, K. O., Sparks, W., Underhill, A. B., Vitagliano, H. D., West, D. K., Macchetto, F., Fitton, B., Barker, P. J., Dunford, E., Gondhalekar, P. M., Hall, J. E., Harrison, V. A. W., Oliver, M. B., Sandford, M. C. W., Vaughan, P. A., Ward, A. K., Anderson, B. E., Boksenberg, A., Coleman, C. I., Snijders, M. A. J., Wilson, R., 1978, The IUE spacecraft and instrumentation, *Nature*, **275**, p. 372.

41. Bohlander, R. A., McMillan, R. W., Gallagher, J. J., 1985, Atmospheric effects on near-millimeter-wave propagation, *IEEE Trans.*, **73**, p. 49.

42. Boksenberg, A., Burgess, D. E., 1972, An image photon counting system for optical astronomy, *Adv. Electron. Electron Phys.*, **33B**, p. 835.

43. Boksenberg, A., Burgess, D. E., 1973, The University College London image photon counting system: performance and observing configuration, *Astronomical Observations with Television-Type Sensors*, ed. Glaspey and Walker, Univ. of British Columbia, Vancouver, p. 21.

44. Bolton, J. G., 1960, Radio telescopes, *Telescopes*, ed. Kuiper and Middlehurst, Univ. of Chicago Press, p. 176.

45. Born, M., Wolf, E., 1975, *Principles of Optics*, Pergamon Press.

46. Borra, E. F., Fletcher, J. M., Poeckert, R., 1981, Multislit photoelectric magnetometer observations of cepheids and supergiants: probable detection of weak magnetic fields, *Astrophys. J.*, **247**, p. 569.

47. Borra, E. F., Vaughan, A. H., 1977, High-resolution polarisation observations of magnetic Ap stars. I. Instrumentation and observations of β Coronae Borealis, *Astrophys. J.*, **216**, p. 462.

48. Borra, E. F., 1976, Polarimetry at the coudé focus: instrumental effects, *Pub. astron. Soc. Pacific*, **88**, p. 548.

49. Boulesteix, J., Georgelin, Y., Marcelin, M., Monnet, G., 1983, First results from cigale scanning Perot–Fabry interferometer, *SPIE*, **445**, p. 37.

50. Boulesteix, J., 1979, First observations of faint extended emission sources with an image photon counting system, *Adv. Electron. Electron Phys.*, **52**, p. 379.

51. Bowen, I. S., 1967, Astronomical optics, *A. Rev. Astron. Astrophys.*, **5**, p. 45.

52. Bowen, I. S., 1964, Telescopes, *Astron. J.*, **69**, p. 816.

53. Bowen, I. S., Vaughan, A. H., 1973, "Nonobjective" gratings, *Pub. astron. Soc. Pacific*, **58**, p. 174.

54. Bowen, I. S., 1960, Schmidt cameras, *Telescopes*, ed. Kuiper and Middlehurst, Univ. of Chicago Press, p. 43.

55. Bowen, I. S., 1960, The 200-inch Hale telescope, *Telescopes*, ed. Kuiper and Middlehurst, Univ. of Chicago Press, p. 1.

56. Bowen, I. S., 1967, Future tools of the astronomer, *Q. J. R. astron. Soc.*, **8**, p. 9.

57. Bowers, F. K., Whyte, D. A., Landecker, T. L., Klinger, R. J., 1973, A digital correlation spectrometer employing multiple-level quantization, *IEEE Trans.*, **61**, p. 1339.

58. Boyd, R. W., 1978, The wavelength dependence of seeing, *J. Opt. Soc. Am.*, **68**, p. 877.

59. Bracewell, R. N., 1979, Computer image processing, *A. Rev. Astron. Astrophys.*, **17**, p. 113.

60. Bracewell, R. N., 1961, Tolerance theory for large antennas, *Inst. Radio Engrs. Trans. Antennas Propagation*, **AP-9**, p. 49.

61. Bracewell, R. N., 1962, Radio astronomy techniques, *Handbuch der Physik*, vol. LIV, ed. Flügge, Springer-Verlag, p. 41.

62. Bracewell, R. N., Thompson, A. R., 1973, The main beam and ring lobes of an east-west rotation-synthesis array, *Astrophys. J.*, **182**, p. 77.

63. Bradt, H. V. D., McClintock, J. E., 1983, The optical counterparts of compact galactic X-ray sources, *A. Rev. Astron. Astrophys.*, **21**, p. 13.

64. Broadfoot, A. L., Kendall, K. R., 1968, The airglow spectrum, 3100–10,000 Å, *J. Geophys. Res.*, **73**, p. 426.

65. Broten, N. W., Clarke, R. W., Legg, T. H., Locke, J. L., McLeish, C. W., Richards, R. S., Yen, J. L., Chisholm, R. M., Galt, J. A., 1967, Diameters of some quasars at a wavelength of 66.9 cm, *Nature*, **216**, p. 44.

66. Broten, N. W., Legg, T. H., Locke, J. L., McLeish, C. W., Richards, R. S., Chisholm, R. M., Gush, H. P., Yen, J. L., Galt, J. A., 1967, Long base line interferometry: a new technique, *Science*, **156**, p. 1592.

67. Brown, R. A., Hilliard, R. L., Phillips, A. L., 1982, Actual blaze angle of the Bausch and Lomb R4 échelle grating, *App. Optics*, **21**, p. 167.

68. Burbidge, G., Burbidge, M., 1967, Quasi-stellar objects, W. H. Freeman.

69. Burke, B. F., 1969, Long-baseline interferometry, *Physics Today*, **22**, p. 54.

70. Campbell, B., 1977, The DDO diode array spectrometer, *Pub. astron. Soc. Pacific*, **89**, p. 728.

71. Campbell, B., Walker, G. A. H., 1979, Precision velocities with an absorption cell, *Pub. astron. Soc. Pacific*, **91**, p. 540.

72. Campbell, B., Walker, G. A. H., 1985, Stellar radial velocities of high precision: techniques and results, *Stellar Radial Velocities*, ed. Davies, A. G. and Latham, D. W., L. Davis Press, p. 5.

73. Campbell, B., Walker, G. A. H., Johnson, R., Lester, T., Yang, S., Auman, J., 1981, Precision radial velocities and residual problems with Reticon arrays, *SPIE*, **290**, p. 215.

74. Cannon, W. H., 1978, The classical analysis of the response of a long baseline radio interferometer, *Geophys. J. R. astron. Soc.*, **53**, p. 503.

75. Carleton, N. P., Hoffmann, W. F., 1978, The multiple-mirror telescope, *Physics Today*, September 1978, p. 30.

76. Carruthers, G. R., Heckathorn, H. M., 1979, Development and application of electrographic image detectors, *Rept. Naval Research Laboratory Prog.*, November 1979, p. 1.

77. Carruthers, G. R., 1973, Apollo 16 far-ultraviolet camera/spectrograph: instrument and operations, *App. Optics*, **12**, p. 2501.

78. Carruthers, G. R., 1979, High resolution large format electronographic cameras for space astronomy, *Adv. Electron. Electron Phys.*, **52**, p. 283.

79. Cayrel, R., Odgers, G. J., 1980, Canada–France–Hawaii Telescope Observers Manual.

80. Chamberlain, J. W., 1961, *Physics of the Airglow*, Academic Press.
81. Charles, M. W., Cook, B. A., 1968, Proportional counter resolution, *Nuc. Instr. Methods*, **61**, p. 31.
82. Chodil, G., Mark, H., Seward, F. D., Swift, C. D., Turiel, I., Hiltner, W. A., Wallerstein, G., Mannery, E. J., 1968, Simultaneous observations of the optical and X-ray spectra of Sco XR-1, *Astrophys. J.*, **154**, p. 645.
83. Choisser, J. P., 1977, Detecting photoelectron images with semiconductor arrays for multichannel photon counting, *Opt. Eng.*, **16**, p. 262.
84. Chou, L. W., 1979, Electron optics of concentric electromagnetic focusing systems, *Adv. Electron. Electron Phys.*, **52**, p. 119.
85. Chow, Y. L., 1972, On designing a supersynthesis antenna array, *IEEE Trans.*, **AP-20**, p. 30.
86. Christiansen, W. N., Högbom, J. A., 1985, *Radiotelescopes*, 2nd edn, Cambridge Univ. Press.
87. Chupp, E. L., 1976, *Gamma Ray Astronomy*, D. Reidel.
88. Church, L., 1978, Sources of a $1/\theta^2$ scattering from optical surfaces, *J. Opt. Soc. Am.*, **68**, p. 1426.
89. Chvojkova, E., 1958, Propagation of radio waves from cosmical sources, *Nature*, **181**, p. 105.
90. Clarke, D., 1974, Polarimetric definitions, *Planets, Stars and Nebulae Studied with Polarimetry*, ed. Gehrels, Univ. of Arizona Press, p. 45.
91. Clarricoats, P. J. B., Poulton, G. T., 1977, High efficiency microwave reflector antennas – a review, *IEEE Trans.*, **65**, p. 1470.
92. Cline, T. L., Desai, U. D., Pizzichini, G., Teegarden, B. J., Evans, W. D., Klebesadel, R. W., Laros, J. G., Hurley, K., Niel, M., Vedrenne, G., Estoolin, I. V., Kouznetsov, A. V., Zenchenko, V. M., Hovestadt, D., Gloeckler, G., 1980, Detection of a fast, intense and unusual gamma-ray transient, *Astrophys. J.*, **237**, p. L1.
93. Cline, T. L., Desai, U. D., Pizzichini, G., Teegarden, B. J., Evans, W. D., Klebesadel, R. W., Laros, J. G., Barat, C., Hurley, K., Niel, M., Vedrenne, G., Estulin, I. V., Mersov, G. A., Zenchenko, V. M., Kurt, V. G., 1981, High-precision location of the 1978 November 19 gamma-ray burst, *Astrophys. J.*, **246**, p. L133.
94. Code, A. D., Davis, J., Bless, R. C., Hanbury Brown, R., 1976, Empirical effective temperatures and bolometric corrections for early-type stars, *Astrophys. J.*, **203**, p. 417.
95. Cohen, M. H., 1969, High-resolution observations of radio sources, *A. Rev. Astron. Astrophys.*, **7**, p. 619.
96. Cohen, M. H., Jauncey, D. L., Kellerman, K. I., Clark, B. G., 1968, Radio interferometry at one-thousandth second of arc, *Science*, **162**, p. 88.
97. Cohen, M. H., Gundermann, E. J., Hardebeck, H. E., Sharp, L. E., 1967, Interplanetary scintillations. II. Observations, *Astrophys. J.*, **147**, p. 449.
98. Cole, T. W., Slee, O. B., 1980, Spectra of interplanetary scintillation, *Nature*, **285**, p. 93, together with 'An alternative explanation', by A. Hewish.
99. Cole, T. W., 1973, An electrooptical radio spectrograph, *IEEE Trans.*, **61**, p. 1321.
100. Coleman, C., 1977, Camera users' guide, IUE Technical Note No. 31, Appleton Lab., Univ. College London.
101. Coles, W. A., Harmon, J. K., 1978, Interplanetary scintillation measurements of the electron density power spectrum in the solar wind, *J. Geophys. Res.*, **83** (A4), p. 1413.

102.   Connes, P., 1970, Astronomical Fourier spectroscopy, *A. Rev. Astron. Astrophys.*, **8**, p. 209.

103.   Conrad, A. G., Irace, W. R., 1983, The first orbiting astronomical infrared telescope system – its development and performance, *SPIE*, **445**, p. 232.

104.   Cooper, B. F. C., 1976, Autocorrelation spectrometers, *Methods of Experimental Physics*, vol. 12, pt. B, ed. Meeks, Academic Press, p. 280.

105.   Coron, N., Danbier, G., Leblance, J., 1971, A new type of helium cooled bolometer, *Infrared Detection Techniques for Space Research*, ed. Manno and Ring, Reidel.

106.   Costain, C. H., Lacey, J. D., Roger, R. S., Large 22-MHz array for radio astronomy, *IEEE Trans.*, **AP-17**, p. 162.

107.   Cromwell, R. H., Angel, J. R. P., 1979, Elimination of corona and related problems with astronomical image tubes, *Adv. Electron. Electron Phys.*, **52**, p. 183.

108.   Currie, D. G., Knapp, S. L., Liewer, K. M., 1974, Four stellar-diameter measurements by a new technique: amplitude interferometry, *Astrophys. J.*, **187**, p. 131.

109.   Dainty, J. C., Scaddan, R. J., 1975, Measurement of the atmospheric transfer function at Mauna Kea, Hawaii, *Mon. Not. R. astron. Soc.*, **170**, p. 519.

110.   Dainty, J. C., Shaw, R., 1974, *Image Science*, Academic Press.

111.   Dainty, J. C., Scaddan, R. J., 1974, A coherence interferometer for direct measurement of the atmospheric transfer function, *Mon. Not. R. astron. Soc.*, **167**, p. 69P.

112.   De Vegt, C., 1978, Current problems of large field astrometry, *Bull. Inf. CDS*, **15**, p. 51.

113.   Dicke, R. H., 1946, The measurement of thermal radiation at microwave frequencies, *Rev. Sci. Instr.*, **17**, p. 268.

114.   Disney, M. J., Sparks, W. B., 1982, On sensible units of apparent flux, *Observatory*, **102**, p. 231.

115.   Ditsler, W. H., 1979, Conversion of a standard television camera to an ISIT acquisition device for astronomy, *SPIE*, **172**, p. 363.

116.   Donaghy, J. F., Canizares, C. R., 1978, The focal plane crystal spectrometer for the HEAO-B satellite, *IEEE Trans.*, **N-S 25**, p. 459.

117.   Dopita, M. A., 1972, Improvement of photomultiplier performance in astronomical applications, *Astrophys. Space Sci.*, **18**, p. 350.

118.   Drake, F. D., 1960, Radio-astronomy radiometers and their calibration, *Telescopes*, ed. Kuiper and Middlehurst, Univ. of Chicago Press, p. 210.

119.   Drake, F. D., 1980, Large telescopes using fixed primaries, *Optical and Infrared Telescopes for the 1990s*, ed. Hewitt, KPNO, p. 649.

120.   Dudinov, V. N., Konichek, V. V., Kuz'menkov, S. G., Tsvetkova, V. S., Rylov, V. S., Gyavgyanen, L. V., Erokhin, V. N., 1982, Speckle interferometry with the BTA telescope, *Instrumentation for Astronomy with Large Telescopes*, ed. Humphries, Reidel, p. 191.

121.   Eichhorn, H., 1974, *Astronomy of Star Positions*, Frank Ungar, New York.

122.   Emerson, D. T., 1976, High resolution observations of neutral hydrogen in M31-II, *Mon. Not. R. astron. Soc.*, **176**, p. 321.

123.   Epps, H. W., Peters, P. J., 1973, Imaging optics for image tubes and television-type sensors, *Astronomical Observations with Television-Type Sensors*, ed. Glaspey and Walker, Univ. of British Columbia, Vancouver, p. 415.

124. Evans, D. E., 1970, Photoelectric measurements of lunar occultations. III. Lunar limb effects, *Astron. J.*, **75**, p. 589.

125. Evans, W. D., Klebesadel, R. W., Laros, J. G., Cline, T. L., Desai, U. D., Pizzichini, G., Teegarden, B. J., Hurley, K., Niel, M., Vedrenne, G., Estoolin, I. V., Kouznetsov, A. V., Zenchenko, V. M., Kurt, V. G., 1980, Location of the gamma-ray transient event of 1979 March 5, *Astrophys. J.*, **237**, p. L7.

126. Everett, P., Hynecek, J., Zucchino, P., Lowrance, J., 1985, Virtual phase charge coupled device imager operated in frontside electron-bombard mode, *Opt. Eng.*, **24**, p. 360.

127. Fahlman, G. G., Glaspey, J. W., Jensen, O., Walker, G. A. H., Auman, J. R., 1974, Attempt to measure stellar magnetic fields using a low-light level television camera, *Planets, Stars and Nebulae Studied with Photopolarimetry*, ed. Gehrels, Univ. of Arizona Press, p. 237.

128. Fahlman, G. G., Glaspey, J. W., 1973, A technique for measuring small displacements in digital spectra, *Astronomical Observations with Television-Type Sensors*, ed. Glaspey and Walker, Univ. of British Columbia, Vancouver, p. 347.

129. Fahlman, G. G., Mochnacki, S. W., Pritchet, C., Condal, A., Walker, G. A. H., 1979, Astronomical imagery with solid-state arrays, *Adv. Electron. Electron Phys.*, **52**, p. 453.

130. Fazio, G. G., 1974, X-ray and gamma-ray detection by means of atmospheric interactions: fluorescence and Čerenkov radiation, *Methods of Experimental Physics*, vol. 12, pt. A Astrophysics, ed. Carleton, Academic Press.

131. Fazio, G. G., 1977, A 102-cm balloon-borne telescope for far-infrared astronomy, *Opt. Eng.*, **16**, p. 551.

132. Federer, C. A., 1964, Some current programmes at Arecibo, *Sky and Telescope*, **28**, p. 2.

133. Fehrenbach, Ch., 1967, Les mesures de vitesses radiales au prisme objectif, *IAU Symposium 30*, ed. Batten and Heard, Academic Press, p. 65.

134. Fernie, J. D., 1976, Note on the dead-time correction in photon counting, *Pub. astron. Soc. Pacific*, **88**, p. 969.

135. Field, G. B., 1982, New instruments for astronomy, *Physics Today*, **35**, p. 25.

136. Findlay, J. W., 1971, Filled aperture antennas for radio astronomy, *A. Rev. Astron. Astrophys.*, **9**, p. 271.

137. Finsen, W. S., 1964, Interferometer observations of binary stars, *Astron. J.*, **69**, p. 319.

138. Fitchel, C., Greisen, K., Kniffen, D., 1975, High-energy gamma-ray astronomy, *Physics Today*, September 1975, p. 42.

139. Fletcher, J. M., Crampton, D., 1973, An astronomical view of high-pressure sodium lamps, *Pub. astron. Soc. Pacific*, **85**, p. 275.

140. Forrest, A. K., 1983, Ultra high resolution radial velocity spectrometer, *SPIE*, **445**, p. 543.

141. Forrest, W. J., Moneti, A., Woodward, C. E., Pipher, J., 1985, The new near-infrared array camera at the University of Rochester, *Pub. astron. Soc. Pacific*, **97**, p. 183.

142. Fouéré, J. C., Lelièvre, G., Lemonier, J. P., Odgers, G. J., Richardson, E. H., Salmon, D. S., 1982, High resolution imagery and slitless spectroscopy at the prime focus of the CFH telescope, *Instrumentation for Astronomy with Large Optical Telescopes*, ed. Humphries, Reidel, p. 143.

143.   Fowler, A., Waddell, P., Mortara, L., 1981, Evaluation of the RCA 512×320 charge-coupled device (CCD) imagers for astronomy, *SPIE*, **290**, p. 34.

144.   Fried, D. L., 1965, Statistics of a geometric representation of wave-front distortion, *J. Opt. Soc. Am.*, **55**, p. 1427.

145.   Gaillardetz, R., Bjorkholm, P., Mastronardi, R., Vanderhill, M., Howland, D., 1978, The HEAO-B Monitor proportional counter instrument, *IEEE Trans.*, *N-S 25*, p. 437.

146.   Galt, J. A., Slater, C. H., Shuter, W. L. H., 1960, An attempt to detect the galactic magnetic field using Zeeman splitting of the hydrogen line, *Mon. Not. R. astron. Soc.*, **120**, p. 187.

147.   Garner, W. V., Johnson, P. G., 1981, Photographic image enhancement of the λ Orionis nebula, *AAS Phot. Bull.*, **27**, p. 10.

148.   Garner, W. V., Meaburn, J., 1979, The combination of unsharp masking and high contrast copying, *AAS Phot. Bull.*, **20**, p. 3.

149.   Garrison, R. F., 1984, *The MK Process and Stellar Classification*, ed. Garrison, Univ. of Toronto Press.

150.   Gascoigne, S. C. B., 1968, Some recent advances in the optics of large telescopes, *Q. J. R. astron. Soc.*, **9**, p. 98.

151.   Geary, J. C., 1979, A floating gate preamplifier design for Reticon (TM) diode arrays, *SPIE*, **172**, p. 82.

152.   Ghausi, M. S., 1965, *Principles and Design of Linear Active Circuits*, McGraw-Hill.

153.   Giacconi, R., Branduardi, G., Briel, U., Epstein, A., Fabricant, D., Feigelson, E., Forman, W., Gorenstein, P., Grindlay, J., Gurskey, H., Harden, F. R., Henry, J. P., Jones, C., Kellog, E., Koch, D., Murray, S., Schreier, F., Seward, F., Tananbaum, H., Topka, K., Van Speybroeck, L., Holt, S. S., Becker, R. H., Boldt, E. A., Serlemitsos, P. J., Clark, G., Canizares, C., Markert, T., Novick, R., Helfland, D., Long, K., 1979, The Einstein (HEAO 2) X-ray observatory, *Astrophys. J.*, **230**, p. 540.

154.   Giacconi, R., Gursky, H., 1974, *X-Ray Astronomy*, Reidel.

155.   Giacconi, R., Kellog, E., Gorenstein, P., Gursky, H., Tananbaum, H., 1971, An X-ray scan of the galactic plane from UHURU, *Astrophys. J.*, **165**, p. L27.

156.   Gillett, F. C., Dereniak, E. L., Joyce, R. R., 1977, Detectors for infrared astronomy, *Opt. Eng.*, **16**, p. 544.

157.   Gillett, F. C., 1980, IR detectors and very large telescopes, *Optical and Infrared Telescopes for the 1990s*, vol. 1, ed. Hewitt, KPNO, p. 293.

158.   Ginaven, R. O., Acton, L. L., Dieball, D. M., Johnson, R. B., Alting-Mees, H. R., Smith, R. D., Beaver, E. A., Harms, R. J., Bartko, F., Flemming, J. C., McCoy, J. G., 1981, Faint object spectrograph (FOS) 512-Digicon detector performance data, *SPIE*, **290**, p. 81.

159.   Goad, J. W., 1982, Kitt Peak National Observatory facilities book.

160.   Gordon, K. J., 1980, The Doppler effect: a consideration of quasar redshifts, *Am. J. Phys.*, **48**, p. 514.

161.   Gordon, M. A., 1976, Computer programs for radio astronomy, *Methods in Experimental Physics*, vol. 12, pt. C, ed. Marton, Academic Press, p. 277.

162.   Gray, P. M., 1983, Fibre optic aperture plate (focap) system at the AAO, *SPIE*, **445**, p. 57.

163.   Green, M., Hansen, J. R., 1969, The application of SEC camera tubes and electrostatic image intensifiers to astronomy, *Adv. Electron. Electron Phys.*, **28B**, p. 807.

164. Greenway, A. H., 1980, Interferometry with arrays of large aperture telescopes, *Optical and Infrared Telescopes for the 1990s*, vol. 2, ed. Hewitt, KPNO, p. 755.

165. Griboval, P. J., 1979, The U.T. electronographic camera: present status, astronomical performance and future developments, *Adv. Electron. Electron Phys.*, **52**, p. 305.

166. Griffin, R. F., 1967, A photoelectric radial velocity spectrometer, *Astrophys. J.*, **148**, p. 465.

167. Gunn, J. E., Westphal, J. A., 1981, Care, feeding, and use of charge-coupled device (CCD) imagers at Palomar Observatory, *SPIE*, **290**, p. 16.

168. Gutiérrez-Moreno, A., Moreno, H., Cortés, G., 1982, A study of atmospheric extinction at Cerro Tololo Inter-American Observatory, *Pub. astron. Soc. Pacific*, **94**, p. 722.

169. Habets, G. M. H. J., Heintze, J. R. W., 1981, Empirical bolometric corrections for the main-sequence, *Astron. Astrophys. Suppl.*, **Ser 46**, p. 193.

170. Habing, H. J., Miley, G., Young, E., Baud, B., Bogess, N., Clagg, P. E., De Jong, T., Harris, S., Raimond, E., Rowan-Robinson, M., Soifer, B. T., 1984, Infrared emission from M31, *Astrophys. J.*, **278**, p. L59.

171. Hachenberg, O., Grahl, B. H., Wielebinski, R., 1973, The 100-meter radio telescope at Effelsberg, *IEEE Trans.*, **61**, p. 1288.

172. Hackwell, J. A., Grasdalen, G. L., Gehrz, R. D., 1982, 10 and 20 micron images of regions of star formation, *Astrophys. J.*, **252**, p. 250.

173. Hall, D. N. B., Ridgway, S., Bell, E. A., Yarborough, J. M., 1979, A 1.4 meter Fourier transform spectrometer for astronomical observations, *SPIE*, **172**, p. 121.

174. Hall, J. A., 1975, Amplifier and amplifier noise considerations, *Solid State Imaging*, ed. Jespers, van de Weile, White, NATO Adv. Study Inst. Ser., Noordhoff, Leyden, p. 535.

175. Hanbury Brown, R., 1974, *The Intensity Interferometer*, Taylor and Francis.

176. Hanbury Brown, R., Twiss, R. Q., 1957, Interferometry of the intensity fluctuations in light. Part 1, *Proc. R. Soc. A*, **242**, p. 300.

177. Hanbury Brown, R., Twiss, R. Q., 1958, Interferometry of the intensity fluctuation in light. Part 3. Applications to astronomy, *Proc. R. Soc. A*, **243**, p. 199.

178. Hanbury Brown, R., 1968, Measurement of stellar diameters, *A. Rev. Astron. Astrophys.*, **6**, p. 13.

179. Hardie, R. H., 1962, Photoelectric reductions, *Astronomical Techniques*, ed. Hiltner, Univ. of Chicago Press, p. 178.

180. Hardy, J. W., Lefèbvre, J. E., Koliopoulos, C. L., 1977, Real-time atmospheric compensation, *J. Opt. Soc. Am.*, **67**, p. 360.

181. Hardy, J. W., 1981, Active optics in astronomy, Proc. ESO Conf. *Scientific Importance of High Angular Resolution at Infrared and Optical Wavelengths*, p. 25.

182. Harmer, C. F. W., 1974, Pupil imagery in astronomical spectrographs, *Mon. Not. R. astron. Soc.*, **167**, p. 311.

183. Hartley, M., McInnes, B., Smith, F. G., 1981, Microthermal fluctuations and their relation to seeing conditions at Roque de los Muchachos Observatory, La Palma, *Q. J. R. astron. Soc.*, **22**, p. 272.

184. Hass, G., 1955, Filmed surfaces for reflecting optics, *J. Opt. Soc. Am.*, **45**, p. 945.

185.   Hauser, M. G., 1982, Space infrared astronomy: overview of NASA planning, *Opt. Eng.*, **21**, p. 758.

186.   Hayes, D. S., Latham, D. W., 1975, A rediscussion of the atmospheric extinction and the absolute spectral-energy distribution of Vega, *Astrophys. J.*, **197**, p. 593.

187.   Hayes, D. S., 1967, An absolute calibration of twelve spectrophotometric standard stars, Thesis, Univ. of California, Los Angeles.

188.   Hazard, C., Mackay, M. B., Shimmins, A. J., 1963, Investigation of the radio source 3C273 by the method of lunar occultations, *Nature*, **197**, p. 1037.

189.   Heacox, W., 1980, An optical fiber spectrograph coupler, *Optical and Infrared Telescopes for the 1990s*, vol. 2, ed. Hewitt, KPNO, p. 702.

190.   Hearn, D. R., Richardson, J. A., Bradt, H. V. D., Clark, G. W., Lewin, W. H. G., Mayer, W. F., McClintock, J. E., Primini, F. A., Rapport, S. A., 1976, MX 1313 + 29: a compact source of very low energy X-rays in Coma Berenices, *Astrophys. J.*, **203**, p. L21.

191.   Hebden, J. C., Morgan, B. L., Vine, H., 1983, Speckle interferometry using a hardwired real-time autocorrelator, *SPIE*, **445**, p. 477.

192.   Heckathorn, H. M., Carruthers, G. R., 1981, An investigation of autoradiographically intensified electrographic imagery, *Pub. astron. Soc. Pacific*, **93**, p. 672.

193.   Hege, E. K., Cocke, W. J., Strittmatter, P. A., Worden, S. P., 1983, High-speed digital processing for speckle interferometry, *SPIE*, **445**, p. 469.

194.   Hege, E. K., Hubbard, E. N., Strittmatter, P. A., 1980, An intensified event-detecting television system for astronomical speckle interferometry, *SPIE*, *264*, p. 29.

195.   Heudier, J. L., 1981, Astronomical photography: its present status, *AAS Phot. Bull.*, **26**, p. 3.

196.   Hewish, A., Burnell, S. J., 1970, Fine structure in radio sources at metre wavelengths – I the observations, *Mon. Not. R. astron. Soc.*, **150**, p. 141.

197.   Hewish, A., 1975, Pulsars and high density physics, *Science*, **188**, p. 1079.

198.   Hickson, P., Fahlman, G. G., Auman, J. R., Walker, G. A. H., Menon, T. K., Ninkov, Z., 1982, CCD photometry of Makarian 421 and 501, *Astrophys. J.*, **258**, p. 53.

199.   Hickson, P., Fahlman, G. G., Walker, G. A. H., 1981, Charge-coupled device (CCD): imaging systems at the University of British Columbia, *SPIE*, **290**, p. 109.

200.   Hill, J. M., Angel, J. R. P., 1983, Optical matching for fiber optic spectroscopy, *SPIE*, **445**, p. 85.

201.   Hill, J., Angel, J. R. P., Lindley, D., 1980, Multiple object spectroscopy, *Optical and Infrared Telescopes for the 1990s*, ed. Hewitt, KPNO, p. 370.

202.   Hoag, A. A., 1976, Sensitometer calibration, *AAS Phot. Bull.*, **13**, p. 14.

203.   Hobbs, L. M., 1969, Interferometric studies of interstellar sodium, *Astrophys. J.*, **157**, p. 135.

204.   Hoffman, M., 1982, The longest darkness, *Observatory*, **102**, p. 208.

205.   Hopkins, H. H., 1950, *Wave Theory of Aberrations*, Oxford Univ. Press.

206.   Houtkamp, J. J., Mulder, H., 1979, Stray light in proximity focused image intensifiers, *Adv. Electron. Electron Phys.*, **52**, p. 159.

207.   Howard, J. N., Burch, D. L., Williams, D., 1955, Near-infrared transmission through synthetic atmospheres, Air Force Cambridge Research Center – Geophysical Research Papers no. 40 (AFCRC-TR-55-213).

208. Humphrey, A., Cabral, R., Brisette, R., Caroll, R., 1978, Imaging proportional X-ray counter for HEAO, *IEEE Trans.*, **N-S 25**, p. 445.

209. Humphries, C. M., Jamar, C., Malaise, D., Wroe, H., 1976, Absolute calibration of the ultraviolet sky survey telescope, *Astron. Astrophys.*, **49**, p. 389.

210. Humphries, C., 1980, Thin mirror telescopes: experience gained with the U.K. infrared telescope, *Optical and Infrared Telescopes for the 1990s*, vol. 2, ed. Hewitt, KPNO, p. 448.

211. Hunten, D. N., 1974, Reshaping and stabilisation of astronomical images, *Methods of Experimental Physics*, vol. 12, pt. A Astrophysics, ed. Carleton, Academic Press, p. 193.

212. Inokuchi, H., Harada, Y., Kondow, T., 1964, Measurement of the intensity of vacuum-ultraviolet light: the application of aromatic hydrocarbons, *J. Opt. Soc. Am.*, **54**, p. 842.

213. Jamar, C., Macau-Hercot, D., Monfils, A., Thompson, G. I., Houziaux, L., Wilson, R., 1976, Ultraviolet bright-star spectrophotometric catalogue, European Space Agency, ESA SR-27.

214. Janesick, J. R., Elliott, T., Collins, S., Marsh, H., Freeman, J., 1984, The future scientific CCD, *SPIE*, **501**, p. 2.

215. Jaquinot, P., Roizen-Dossier, B., 1964, Apodisation, *Progress in Optics III*, ed. Wolf, North-Holland, p. 31.

216. Jaquinot, P., 1960, New developments in interference spectroscopy, *Rep. Prog. Phys.*, **23**, p. 267.

217. Jelley, J. V., 1958, Čerenkov radiation and its applications, Pergamon Press.

218. Jennison, R. C., Das Gupta, M. K., 1956, The measurement of the angular diameters of two intense radio sources, *Phil. Mag.*, Ser. (8), **1**, p. 66.

219. Jennison, R. C., 1958, A phase sensitive interferometer technique for the measurement of the Fourier transforms of spatial brightness distributions of small angular extent, *Mon. Not. R. astron. Soc.*, **118**, p. 276.

220. Johnson, H. J., 1980, The absolute calibration of stellar spectrophotometry, *Rev. Mex. Ast. Af.*, **5**, p. 25.

221. Johnson, M. A., Betz, A. L., Townes, C. H., 1974, 10 μm heterodyne stellar interferometer, *Phys. Rev. Lett.*, **33**, p. 1617.

222. Jones, R. C., 1968, How images are detected, *Sci. Am.*, **219**, p. 111.

223. Jones, R. V., Richards, J. C., 1954, The polarisation of light by narrow slits, *Proc. R. Soc. A*, **225**, p. 122.

224. Jouyce, R. M., Becker, R. H., Birsa, F. B., Holt, S. S., Noordzy, M. P., 1978, The Goddard Space Flight Center solid state spectrometer for the HEAO-B mission, *IEEE Trans.*, **N-S 25**, p. 453.

225. Kellog, E., Henry, P., Murray, S., Van Speybroeck, L., 1976, High resolution imaging X-ray detector, *Rev. Sci. Instr.*, **47**, p. 282.

226. KenKnight, C. E., 1977, Methods of detecting extrasolar planets I. Imaging, *Icarus*, **30**, p. 422.

227. Kharitonov, A. V., Tereshchenko, V. M., Knyazeva, L. N., Boiko, P. N., 1980, Calibration of the spectra of selected stars. I. Vega, *Soviet Astron.*, **24**(2), p. 168.

228. King, I. R., 1971, The profile of a star image, *Pub. astron. Soc. Pacific*, **83**, p. 199.

229. Kingston, R. H., 1978, *Detection of Optical and Infrared Radiation*, Springer-Verlag.

230.   Klebesadel, R. W., Strong, I. B., Olsen, R. A., 1973, Observations of gamma-ray bursts of cosmic origin, *Astrophys. J.*, **182**, p. L85.

231.   Klemperer, W. K., 1972, Long-baseline radio interferometry with independent frequency standards, *IEEE Trans.*, **60**, p. 602.

232.   Koo, D. C., 1985, Optical multicolors: a poor person's *z* machine for galaxies, *Astron. J.*, **90**, p. 418.

233.   Korff, D., 1973, Analysis of a method for obtaining near-diffraction limited information in the presence of atmospheric turbulence, *J. Opt. Soc. Am.*, **63**, p. 971.

234.   Kormendy, J., 1973, Calibration of direct photographs using brightness profiles of field stars, *Astron. J.*, **78**, p. 255.

235.   Kraus, J. D., 1966, *Radio Astronomy*, McGraw-Hill.

236.   Kron, G. E., 1974, Electronographic tubes, *Methods of Experimental Physics*, vol. 12, pt. A Astrophysics, ed. Carleton, Academic Press, p. 252.

237.   Kruse, P. W., McGlauchlin, L. D., McQuistan, R. B., *Elements of Infrared Technology: Generation, Transmission, and Detection*, Wiley.

238.   Kubierschky, K., Austin, G. K., Harrison, D. C., Roy, A. G., 1978, The high resolution imaging instrument for HEAO-B, *IEEE Trans.*, **N-S 25**, p. 430.

239.   Labeyrie, A., Bonneau, D., Stachnik, R. V., Gezari, D. Y., 1974, Speckle interferometry. III. High-resolution measurements of twelve close binary systems, *Astrophys. J.*, **194**, p. L147.

240.   Labeyrie, A., 1978, Stellar interferometry methods, *A. Rev. Astron. Astrophys.*, **16**, p. 77.

241.   Labeyrie, A., 1975, Interference fringes obtained on Vega with two optical telescopes, *Astrophys. J.*, **196**, p. L71.

242.   Labeyrie, A., 1980, Interferometry with arrays of large-aperture ground based telescopes, *Optical and Infrared Telescopes for the 1990s*, vol. 2, ed. Hewitt, KPNO, p. 786.

243.   Lampton, M., Carlson, C. W., 1979, Low distortion resistive anodes for two dimensional position sensitive MCP systems, *Rev. Sci. Instr.*, **50**, p. 1093.

244.   Lampton, M., Cash, W., Malina, R. F., Boyer, S., 1977, Design, fabrication and performance of two grazing incidence telescopes for celestial extreme ultraviolet astronomy, *SPIE*, **106**, p. 93.

245.   Lampton, M., 1981, The microchannel image intensifier, *Sci. Am.*, **245**, p. 62.

246.   Lang, K. R., 1980, *Astrophysical Formulae*, Springer-Verlag.

247.   Langford, D. L., Simmonds, J. J., Ozawa, T., Long, E. C., Paris, R., 1983, Infrared and visible detector electronics for the Infrared Astronomical Satellite (IRAS), *SPIE*, **445**, p. 244.

248.   Latham, D. W., 1974, Detective performance of photographic plates, *Methods of Experimental Physics*, vol. 12, pt. A Astrophysics, ed. Carleton, Academic Press, p. 221.

249.   Latham, D. W., 1976, The effects of push development and hydrogen sensitization on the detective performance of Kodak plates, types IIa and 127-04, *AAS Phot. Bull.*, **13**, p. 9.

250.   Leach, R. W., Gursky, H., 1979, The cosmic ray background in charge coupled devices, *Pub. astron. Soc. Pacific*, **91**, p. 855.

251.   Lemaître, G., 1980, Elastic relaxation figuring for mirror mass production, 1980, *Optical and Infrared Telescopes for the 1990s*, vol. 2, ed. Hewitt, KPNO, p. 896.

252.   Lemke, D., Klipping, G., Grewing, M., Preussner, P., Martin, W., Trinks, H., Drapatz, S., Hofman, R., Hartman, G., Proetel, K., 1980, The German